CHARLES, THE ALTERNATIVE KING

AN *UN*AUTHORISED BIOGRAPHY

REVISED EDITION

Edzard Ernst

SOCIETAS
essays in political
& cultural criticism

imprint-academic.com

Copyright © Edzard Ernst, 2023

The moral rights of the author have been asserted.
No part of this publication may be reproduced in any form without permission, except for the quotation of brief passages in criticism and discussion.

1st edition published 2022

Published in the UK by
Imprint Academic Ltd., PO Box 200, Exeter EX5 5YX, UK

Distributed in the USA by
Ingram Book Company,
One Ingram Blvd., La Vergne, TN 37086, USA

ISBN 9781788361064 paperback

A CIP catalogue record for this book is available from the British Library and US Library of Congress

Contents

	Foreword by Nick Ross	v
	Preface to the Second Edition	ix
1.	Why this Book?	1
2.	Why this Author?	5
3.	Words and Meanings	10
4.	How Did It All Start?	13
5.	Laurens van der Post	17
6.	The British Medical Association	25
7.	Talking Health	31
8.	Osteopathy	37
9.	Chiropractic	43
10.	The Foundation of Integrated Health	50
11.	Open Letter to *The Times*	56
12.	The Model Hospital	62
13.	Integrated Medicine	66
14.	The Gerson Therapy	73
15.	Herbal Medicine	77
16.	The Smallwood Report	82
17.	World Health Organisation	90

18. Traditional Chinese Medicine	96
19. The 'GetWellUK' Study	100
20. Bravewell	106
21. Duchy Originals Detox Tincture	110
22. Charles' Letters to Health Politicians	115
23. The College of Medicine and Integrated Health	120
24. The Enemy of Enlightenment	126
25. Harmony	132
26. Antibiotic Overuse	142
27. Ayurvedic Medicine	147
28. Social Prescribing	154
29. Homeopathy	160
30. Final Thoughts	169
31. Long Live the King!	180
Glossary	185
End Notes	192
Index	209

Nick Ross

Foreword

Nothing like this has been written before. The British royal family is a source of global fascination and King Charles, first as heir to the throne, now as the king, has been subject to thousands of headlines and millions of words of appraisal, tittle-tattle, praise, and reproach. But no one has systematically reviewed his public and passionate views on alternative medicine and subjected them to scrutiny as this book does. This is not a conventional biography and nor does it concern itself with monarchy as such; and yet it says a great deal about the king.

I believe Charles is a fundamentally good human being. That said, I won't duck the elephant in the room: I had several times met and liked the former Princess of Wales and so instinctively took Diana's side during the painful breakup and its aftermath; but I also understood that marriage and family life in the gilded cage of royalty was not easy. I disliked Charles' sense of entitlement in his evangelical forays into the politics of architecture and the environment; yet I admired his courage, shared his interests, and even agreed with some of his judgments. I was astonished when he disparaged rationality (saying he was proud to be 'an enemy of the Enlightenment'), but assumed he was quoted out of context or that excitement had got the better of him. I winced when his helicopter clattered over my office just as I was reading of his passion for cutting carbon emissions, but we all find it hard to live up to our own pieties. Even when he is wrong, I believe his motives are right—and it is far better to err in that direction than in the opposite one.

Indeed, if it is not impertinent to make comparisons, I share some features with him (and, as it happens, we both share some with the author of this book). Like the King I was a post-war baby, have some German blood, and was brought up to believe in unscientific theories. Indeed, like His Majesty I was once a fan of the Jungian and spiritualist writer Laurens van der Post, whose charisma veiled his carelessness about the truth, and who (as this book relates) came to have such a profound impact on the former Prince of Wales. Where I diverged from Charles' outlook was at university. I had gone to study psychology, assuming Freudian insights would allow me to peer into people's souls. But it turned out that Freud was simply the most colourful of many theorists of the mind, and that the strength of his opinions was in inverse proportion to any evidence to back them up. Psychoanalysis was more like a religion than a science. Rather to my alarm (for I was terrible at science at school), I was inducted into a world where beliefs had to be verified by evidence—indeed, where experiments started with the so-called null hypothesis, which means starting with the expectation that you are wrong. I discovered the humility of science—that while faith puts ideas on a pedestal and bows before them, in contrast science (at least good science) throws stones at its treasured ideas and tries to knock them off their plinth. It is that scepticism and readiness to be change one's mind in the face of improved knowledge which transformed life expectancy, led to a rising cadence of human understanding and achievement, and which inspired agricultural, industrial, artistic, social, and information revolutions.

Perhaps, above all, I learned how seduction by easy thinking—intuition and personal experience (one's own, or anecdotes from others)—can be fatally misleading—and held back all of those advances.

So while the then Prince Charles continued to follow the teachings of van der Post, not least championing anecdotal medicine, I helped found a charity called Healthwatch devoted to exposing quackery and pseudoscience. And that is where I came across Professor Ernst.

Here was a disciple of alternative medicine who subjected his beliefs to the acid tests of controlled trials and systematic reviews, and who, as the results came in, had the modesty to accept that his faith in folk therapies was substantially misplaced. As the world's first professor of complementary and alternative medicine (or CAM), Edzard became the foremost authority on what worked and what didn't in the sphere of anecdotal therapy, and it turned out that many treatments didn't work as their disciples (or their sales blurb) claimed.

The fact is that most ailments resolve of their own accord, albeit sometimes helped by sleight of hand — the so-called placebo effect. Indeed, even in these days of evidence-based medicine there is still some truth to Voltaire's jibe that the art of a great physician is to keep his patient amused until she either gets better or dies. We are all inclined to believe that B results from A — that if we take a pill and the symptoms subside then the pill must have been effective. And just as gamblers remember their wins and forget their losses, so we are inclined to believe what we are inclined to believe. CAM is intuitive and intrinsically attractive. For many people it becomes an article of faith. For others, some avaricious but mostly sincere, it becomes a career and a source of income.

The idea that a trained homeopathist like Ernst should challenge that faith was, for many, tantamount to betrayal. An informal campaign developed, with supporters of Prince Charles at the fore, seeking to close down his academic department — and as funds dried up, Exeter University found it expedient to withdraw support. The work there has now been abandoned and the researchers have scattered. Nor did Prof Ernst always win friends among conventional doctors, some of whom resented the fairmindedness with which he acknowledged there are 'hidden gems' among fringe remedies, and that pharmaceutical companies have sometimes embroidered their evidence too.

The ferocity of the attacks on him was such that along with the likes of Anthony Fauci he was awarded the John Maddox Prize by *Nature* and Sense About Science for standing up for evidence in the face of hostile disbelievers.

King Charles has endured savage critics too, not least those who attack him for what for most mortals would be private family matters; but including architects who deride his oldfashioned aesthetics, and scientists who deplore his unwillingness to explore facts which undermine his own philosophies. He has borne a lifetime of frustration, always in the public eye and at the apex of the constitution; yet kept for a lifetime at the margins. He must reconcile his regal stiffness with the age of informality, and juggle the contradictions of being (at the time) His Royal Highness when deference has become a dirty word.

So these are both people of substance, both used to insult and abuse, and who for very different reasons (one from experience and mystic values, and one from data analysis and experiment) epitomise contrasting ways of judging what can be trusted and what can't.

The fact that you have picked up this book and got this far suggests you have an interest in the subject—and that means you may well have strong views of your own. If so, I urge you to set them aside. Neither King Charles nor Professor Ernst are incarnations of evil, as their detractors sometimes paint them. Both care about the fate of people who are worried sick, and for that matter both are also concerned about the worried well. The struggle between the professor and the king is not a competition about who has most compassion. It is a contest between honest belief and honest evidence.

© Nick Ross 2021

Edzard Ernst

Preface to the Second Edition

When Queen Elizabeth II, Charles' mother, passed away on 8 September 2022, Charles acceded to the throne. Up to that point, he had been the longest-serving heir apparent in British history and, at the age of 74, he was the oldest person ever to assume the throne.

Now that Charles is King of the United Kingdom and 14 other Commonwealth realms, the original title of my biography seems no longer appropriate. We have therefore decided to publish the second edition under a slightly altered title: *Charles, The Alternative King*.

In 1660, Charles II founded the Royal Society, the world's first and foremost academy of science with the motto *'nullius in verba'*. He was an enthusiastic champion of the Enlightenment. Even today, the 'King Charles II Medal' is awarded to foreign Heads of State or Government who have made outstanding contributions to furthering scientific research in their country. Some 400 years later, Charles III is a patron of this auspicious society, yet he also once stated that he is proud of having been called an enemy of the Enlightenment. The contradiction has only become more obvious now that Charles has become King.

When Queen Elizabeth died, it signalled the end of an era. Will the new Carolean age be one where tree-huggers and other 'irrationalists' feel empowered by a like-minded sovereign? The first edition of this biography concluded by predicting that, as King, Charles will stop openly campaigning for alternative medicine, not least because he once said so himself. I have no reason to doubt that this is true. At the same time, we cannot deny that, as King, his influence at home and abroad has grown immeasurably.

This book chronicles Charles' track record in promoting pseudo- and anti-science in the realm of alternative medicine. The new edition includes an additional final chapter with a summary of some of the scientific evidence that has emerged since this biography was first published. It demonstrates that the concerns about the safety and efficacy of the treatments in question are becoming ever more disquieting. Whether such data will tame the alternative bee under the royal bonnet seems, however, doubtful.

But who knows, the transition from Prince of Wales to King is unquestionably a momentous step; let's hope that it prompts a fundamental conversion of Charles into a monarch of rational thought and advocate of rigorous science, but—'*nullius in verba*'—take nobody's word for it.

Edzard Ernst
Cambridge, January 2023

One

Why this Book?

Over the past two decades, I have supported efforts to focus healthcare on the particular needs of the individual patient, employing the best and most appropriate forms of treatment from both orthodox and complementary medicine in a more integrated way.[1]
— *The Prince of Wales 1997*

This is a charmingly British understatement, indeed! Charles has been the most persistent champion of alternative medicine in the UK and perhaps even in the world. Since the early 1980s, he has done everything in his power

- to boost the image of alternative medicine,
- to improve the status of alternative practitioners,
- to make alternative therapies more available to the general public,
- to lobby that it should be paid for by the National Health Service (NHS),
- to ensure the press reported favourably about the subject,
- to influence politicians to provide more support for alternative medicine.

He has fought for these aims on a personal, emotional, political, and societal level. He has used his time, his intuition, his influence, and occasionally his money to achieve his goals. In 2010, he even wrote a book, *Harmony*, in which he explains his ideas in some detail[2] (discussed in Chapter 25, arguably the central chapter of this biography). Charles has thus become the undisputed champion of the realm of alternative medicine. For that he is admired by alternative practitioners across the globe.

Yet, his relentless efforts are not appreciated by everyone (another British understatement!). There are those who view his interventions as counterproductive distractions from the important and never-ending task to improve modern healthcare. There are those who warn that integrating treatments of dubious validity into our medical routine will render healthcare less efficient. There are those who claim that the King's preoccupation with matters that he is not qualified to fully comprehend is a disservice to public health. And there are those who insist that the role of the heir to the throne does not include interfering with health politics.

- So, are Charles' ideas new and exciting?
- Or are they obsolete and irrational?
- Has Charles become the saviour of UK healthcare?
- Or has he hindered progress?
- Is he a role model for medical innovators?
- Or the laughing stock of the experts?
- Is he a successful reformer of healthcare?
- Or are his concepts doomed to failure?

Charles appears to evade critical questions of this nature. Relying on his intuition, he unwaveringly pursues and promotes his personal beliefs, regardless of the evidence (Box 1). He believes strongly in his mission and is, as most observers agree, full of good intentions. If he even notices any criticism, it is merely to reaffirm his resolve and redouble his efforts. He is reported to work tirelessly, and one could easily get the impression that he is obsessed with his idea of integrating alternative medicine into conventional healthcare.

I have observed Charles' efforts around alternative medicine for the last 30 years. Occasionally, I was involved in some of them. For 19 years, I have headed the world's most productive team of researchers in alternative medicine. This background puts me in a unique position to write this account of Charles' 'love affair' with alternative medicine. It is not just a simple outline of Charles' views and actions but also a critical analysis of the evidence that does or does not support them. In writing it, I pursue several aims:

1. I want to summarise this part of medical history, as it amounts to an important contribution to the recent development of alternative medicine in the UK and beyond.
2. I hope to explain how Charles and other enthusiasts of alternative medicine think, what motivates them, and what logic they follow.
3. I will contrast Charles' beliefs with the published evidence as it pertains to each of the alternative modalities (treatments and diagnostic methods) he supports.
4. I want to stimulate my readers' abilities to think critically about health in general and alternative medicine in particular.

My book will thus provide an opportunity to weigh the arguments for and against alternative medicine. In that way, it might even provide Charles with a substitute for a discussion about his thoughts on alternative medicine which, during almost half a century, he so studiously managed to avoid.

In pursuing these aims there are also issues that I hope to avoid. From the start, I should declare an interest. Charles and I once shared a similar enthusiasm for alternative medicine. But, as new evidence emerged, I changed my mind and he did not. This led to much-publicised tensions and conflicts. Yet it would be too easy to dismiss this book as an act of vengeance. It isn't. I have tried hard to be objective and dispassionate, setting out Charles' claims as fairly as I can and comparing them with the most reliable evidence. As much as possible:

1. I do <u>not</u> want my personal discord with Charles to get in the way of objectivity.
2. I do <u>not</u> want to be unfairly dismissive about Charles and his ambitions.
3. I do <u>not</u> want to be disrespectful about anyone's deeply felt convictions.
4. I do <u>not</u> aim to weaken the standing of our royal family.

My book follows Charles' activities in roughly chronological order. Each time we encounter a new type of alternative medicine,

I will try to contrast Charles' perceptions with the scientific evidence that was available at the time. Most chapters of this book are thus divided into four parts:

1. A short introduction.
2. Charles' views.
3. An outline of the evidence.
4. A comment about the consequences.

While writing this book, one question occurred to me regularly: Why has nobody so far written a detailed history of Charles' passion of alternative medicine? Surely, the account of Charles' 'love affair' with alternative medicine is fascinating, diverse, revealing, and important!

I hope you agree.[1]

BOX 1

The nature of evidence in medicine and science

- Evidence is the body of facts, often created through experiments under controlled conditions, that lead to a given conclusion.
- Evidence must be neutral and give equal weight to data that fail to conform to our expectations.
- Evidence is normally used towards rejecting or supporting a hypothesis.
- In alternative medicine, the most relevant hypotheses often relate to the efficacy of a therapy.
- Such hypotheses are best tested with controlled clinical trials where a group of patients is divided into two subgroups and only one is given the therapy to be tested; subsequently the results of both groups are compared.
- Experience does not amount to evidence and is a poor indicator of efficacy; it can be influenced by several phenomena, e.g. placebo effects, natural history of the condition, regression towards the mean.
- If the results of clinical studies are contradictory, the best available evidence is usually a systematic review of the totality of rigorous trials.
- Systematic reviews are methods to minimise random and selection biases. The most reliable systematic reviews are, according to a broad consensus, those from the Cochrane Collaboration.

---

[1] I am grateful to Richard Rasker for correcting linguistic errors etc.

Two

Why this Author?

There is no shortage of biographies of King Charles. Most of the authors are journalists; I am a physician and a scientist.

Why does that matter?

All previous biographers covered Charles' entire life—that was their aim. My book is fundamentally different; it is focused on just one single aspect of Charles' activities: alternative medicine. All the previous authors have, of course, mentioned this subject as well, yet they were not in an ideal position to do it justice. For that task, they would have needed a different set of skills. Journalists are not scientists. Their biographies are broader, mine is narrow and goes into more detail on one specific subject.

Previous authors reported Charles' views on alternative medicine much like they discussed his thoughts on architecture. To them, this made sense: both topics are similarly controversial; on both subjects, Charles lacks professional competence; and both are close to his heart. But there is nevertheless an important difference. When Charles proclaims that he prefers Georgian to modern styles of architecture, we can agree or disagree with him. It is largely a matter of taste.

When Charles says he believes that iridology is a valuable diagnostic technique (Chapter 25) or that the Gerson therapy is worth considering as a treatment for cancer (Chapter 14), he expresses his opinion in much the same way as he expresses his views on architecture. But architecture is not medicine. In medicine, we have evidence about what works and what not, about what is safe and what is dangerous. Opinion does not amount to evidence; often it is the exact opposite. You might not like a

Georgian house, but you can live in it; it is fit for that purpose. If a cancer patient opts to use the Gerson therapy to cure her disease, she will hasten her death. The Gerson therapy is not fit for the purpose of curing cancer. Everybody is allowed to have their own opinions, but nobody is allowed to have their own facts. And no amount of belief generates a fact.

When journalists cover Charles' history in relation to alternative medicine, they mainly report about his beliefs, his opinions, his preferences, his persuasions, his ambitions, etc. And that is, of course, absolutely fine; that's what journalists do, and that's what they should do. However, for deciding what is right and wrong in medicine, we need more. And authors of biographies cannot easily provide that extra requirement because they do not possess the expertise needed to understand research methodology and medical evidence. As excellent as some of these biographies are, they cannot authoritatively present the evidence which is necessary to put Charles' opinions on alternative medicine in an evidence-based perspective.

One of my favourite biographies of Charles is the one by Catherine Mayer. In it she suggests that Charles and I have more in common than meets the eye.[3] She might be right:

- We were born in the same year.
- We both have German blood.
- We both had/have powerful mothers who delegated much of our upbringing to others.
- We both disliked being at boarding school.
- We both were not exactly brilliant at school nor at university.
- We both grew to be slightly introverted, insecure adolescents.
- We both were brought up with alternative medicine.
- We both believed in the benefits of some alternative treatments.
- We both can be stubbornly determined.

Homeopathy and other forms of alternative medicine were never unusual to me. Our family doctor was a homeopath. A clear distinction between alternative and conventional healthcare was made only when I went to medical school. After I graduated, my first job as a junior doctor was in a homeopathic hospital. Later I occasionally used homeopathy and other alternative medicines as a clinician. When I became head of a large clinical department at the medical school in Vienna, I employed several alternative therapies and conducted clinical trials of acupuncture, homeopathy, and several other alternative therapies.

In 1993, I was appointed chair of complementary medicine at the University of Exeter. My remit was to scientifically investigate all aspects of alternative medicine. I was convinced that alternative medicine had considerable potential. Yet, I also knew that, as a scientist, I had to leave all personal baggage behind and try to be as objective as possible. What counted was not the direction of the results of our findings but the reliability of the research. As a scientist, I had no intention to promote this or that alternative therapy, my aim was to find out which treatments worked for which conditions and what risks they entailed.

At Exeter, I built up a multidisciplinary team of about 20 researchers. Essentially, we tried to find out which alternative modalities generate more good than harm. Many believers in alternative medicine, including perhaps Charles, found this agenda bewildering, but for me it seemed the only possible strategy to pursue. The questions of efficacy and safety are so fundamental, I felt (and still feel), that all other issues seem trivial in comparison.

For 20 years, we conducted research into homeopathy, acupuncture, chiropractic, herbal medicine, and many other alternative treatments (Box 2). Together, we published more scientific papers than any other team researching this area. According to a 2020 analysis by John Ioannidis *et al.* entitled 'Updated Science-Wide Author Databases of Standardized Citation Indicators',[4] I am ranked:

- No. 107 amongst the 160,000 most-cited scientists of all disciplines worldwide.
- No. 1 amongst all researchers in the category of 'Complementary & Alternative Medicine'.

- No. 1 amongst all researchers from the University of Exeter.
- No. 11 amongst all scientists from the UK.

It is thus true to say that:

- I experienced alternative medicine as a patient.
- I practised alternative medicine as a clinician.
- I researched alternative medicine as a scientist.

Charles was certainly aware of the new post at Exeter; he was peripherally involved in creating it (Chapter 7), and he even asked to see a copy of my inaugural lecture.[5] Later, he invited me to St James's Palace where I met him personally for the first time. When I received an invitation to Highgrove, we met a second time. During these two encounters, there never was an occasion for an exchange of views. Yet, he knew of my work — I am sure of this, because he cited some of it in his speeches — and, of course, I closely followed his activities.

The fact that our interests differed is undeniable:

- He was promoting alternative medicine, while I was testing it.
- He wanted integration of alternative medicine, while I cautioned that integrating untested treatments into routine medical practice risked being counterproductive or even dangerous (Chapter 13).

These differences occasionally caused controversies. These widely publicised disputes are, however, irrelevant for this book; I will therefore merely mention them without going into details (those interested can find a full account in my memoir, *A Scientist in Wonderland*[5]).

My stance, knowledge, and experience as a scientist are unquestionably different from Charles' views, preferences, and ambitions. Yet, I think my professional background is well suited to provide an evidence-based analysis of Charles' passion for alternative medicine.

BOX 2

Examples of alternative modalities on which my Exeter team published research papers

- Acupressure
- Acupuncture
- Anthroposophical medicine
- Aromatherapy
- Autogenic training
- Autologous blood therapy
- Ayurveda
- Bach flower remedies
- Biofeedback
- Breathing techniques
- Chelation therapy
- Chiropractic
- Colonic irrigation
- Cupping
- Exercise
- Feldenkrais method
- Flotation therapy
- Guided imagery
- Healing
- Herbal medicine
- Homeopathy
- Hypnotherapy
- Iridology
- Kombucha
- Magnets
- Massage
- Meditation
- Moxibustion
- Neural therapy
- Osteopathy
- Qigong
- Reflexology
- Reiki
- Singing exercises for snoring
- Spinal manipulation
- Supplements
- Tai Chi
- Yoga

Three

Words and Meanings

3.1. Why call it 'biography'?

A biography is a written account of a person's life. Most of the published biographies of Charles are informative, some are a bit too devout for my taste, and some are full of regrettable errors. They all have in common that they try to be complete by covering most aspects of Charles' life. Thus, they also mention Charles' passion for alternative medicine. But none has a focus on this topic.

My book is solely about Charles' activities related to alternative medicine and it omits almost all other aspects. One could therefore argue that it is not a biography at all. However, because alternative medicine has been a central theme in Charles' life that has occupied him during most of his life, I feel it is justified to call my book a 'biography'. It is of course not a typical biography and it is therefore not in competition with the any of the existing ones. But it is an account of one important theme that dominated much of Charles' adult life.

3.2. Why call it 'unauthorised'

An authorised biography is written with the cooperation of the person whom the book is about. By contrast, an unauthorised biography is one written without such assistance. Most of the existing biographies of Charles are not authorised (in fact, I know only one that is[6]). Yet the authors do not normally point this out on the book cover. Most of these books are more or less flattering to Charles. Mine isn't all that flattering, I am afraid. It is a critical evaluation of Charles' activities in the area of alternative medicine. I am fairly sure that Charles would not have cooperated with such a book, regardless of who the author might be. I therefore think it is only fair to point out from the very outset that my biography is unauthorised.

3.3. Why call it 'alternative medicine'?

In this book, I frequently use the umbrella term 'alternative medicine' for the area of healthcare we are discussing. The reason for this choice of terminology is simple: it is the word that the public knows best. It also is the name that Charles employs regularly. But this is not to say that it is the only or even the best word for it. There are many other terms with similar meanings[7] (Box 3).

Of course, the umbrella terms listed in Box 3 are not entirely synonymous but all have subtly different meanings. However, they do overlap a great deal and are often used in parallel for the same modalities. For instance, homeopathy is often used in addition to conventional medicine, which would make it a complementary therapy. Yet, Samuel Hahnemann, the founding father of homeopathy, called people who did that 'traitors' and strictly forbade combining it with other forms of healthcare.[8] Therefore, homeopathy is strictly speaking an alternative therapy.

In his speeches and articles, Charles uses several different terms, including alternative, complementary, and integrated medicine, usually without making clear distinctions between them. Such Babylonian confusion might irritate the public (and the readers of this book). In order to make things not more perplexing, I will follow Charles' lead and employ the terminology that he happens to use at any given moment — and more often than not, this is 'alternative medicine'.

Box 3

Different terminologies

- Fringe medicine is rarely used today.[9] It denotes the fact that the treatments under this umbrella are not in the mainstream of healthcare. Most advocates would find the word derogatory, and therefore it is now all but abandoned.
- Unorthodox medicine is a fairly neutral term describing the fact that medical orthodoxy tends to shun most of the treatments under this umbrella. Strictly speaking, the word is incorrect; the correct term would be 'heterodox medicine', a term that is rarely used.
- Unconventional is also a neutral term but it is wide open to misunderstandings: any innovation in medicine might initially be called unconventional.
- Traditional medicine describes the fact that most of the modalities in question have been used for centuries and thus have a long tradition of usage. However, as the term is also sometimes used for conventional medicine, it is confusing and far from ideal.
- Alternative medicine is the term everyone seems to know and which is therefore most commonly employed in non-scientific contexts. In the late 1980s, some experts pointed out that the word could give the wrong impression: most of the treatments in question are not used as a replacement of but as an adjunct to conventional medicine.
- Complementary medicine became subsequently popular based on the above consideration. It accounts for the fact that the treatments tend to be used by patients in parallel with conventional medicine.
- Complementary and alternative medicine (CAM) describes the phenomenon that many of the treatments can be employed either as a replacement of or as an adjunct to conventional medicine.
- Holistic medicine denotes the fact that practitioners often pride themselves on looking after the whole patient — body, mind, and spirit. This could lead to the erroneous impression that conventional clinicians do not aim to practice holistically. Yet, good healthcare always has been holistic.[10] Therefore, the term can be misleading.
- Natural medicine describes the notion that many of the methods in question are natural. The term seems attractive. However, any close analysis will show that many of the treatments in question are not truly natural.[8] Therefore, this term too is misleading.
- Integrated medicine is currently popular and much used by Charles. As we will see in Chapter 13, the term is nevertheless problematic.
- Integrative medicine is the word used in the US for integrated medicine.
- So-called alternative medicine (SCAM) is a term I often employ these days.[11] It accounts for two important facts: 1) if a treatment does not work, it cannot possibly serve as an adequate alternative; 2) if a therapy does work, it should be part of conventional medicine. Thus, there cannot be an alternative medicine, as much as there cannot be an alternative chemistry or physics. Some advocates find the term and its abbreviation derogatory. Yet intriguingly, my decision to use this term in most of my current writings was inspired by Charles: in his book *Harmony*, he repeatedly speaks of 'so-called alternative treatments'.[2]

Four

How Did It All Start?

Charles was born at Buckingham Palace on 14 November 1948. For a member of the British royal family, he had a relatively normal, almost uneventful childhood.

- He attended two of his father's former schools, Cheam Preparatory School and, from 1962 to 1967, Gordonstoun in Scotland. He left with 6 GCE O-levels and two A-levels (History and French, at grades B and C).
- In October 1967, he was admitted to Trinity College in Cambridge, where he read Anthropology, Archaeology, and History.
- During his second year, Charles attended the University College of Wales in Aberystwyth, studying Welsh History and Language for one term.
- In 1970, he graduated from Cambridge with a Bachelor of Arts, the first heir apparent to earn a university degree.
- In 1958, Charles was created Prince of Wales and Earl of Chester.
- He took his seat in the House of Lords in 1970.
- From the mid-1970s, Charles began to take on more public duties; for instance, by founding The Prince's Trust.
- Charles had several well-publicised affairs.
- Until 1992, Charles was an avid player of competitive polo.
- In 1981, Charles married Lady Diana Spencer.
- Charles and Diana divorced on 28 August 1996.
- In 2005, Charles married Camilla Parker Bowles.

- Charles was the longest-serving Prince of Wales in history.
- Now he has become king, he is the oldest person ever to do so.

But you know all this — and if you don't, there are plenty of good biographies to inform you about a plethora of fascinating details of Charles' life. As mentioned before, this book is not a conventional biography, it is an analysis of his passion for alternative medicine.

So, how did Charles' interest in alternative medicine start?

Charles was 'easily cowed by the forceful personality of his father', whose rebukes for 'a deficiency in behaviour or attitude... easily drew tears', notes Jonathan Dimbleby, adding that the late Prince Philip was 'well-meaning but unimaginative'. For his mother, too, Charles finds rather harsh words; she was 'not indifferent so much as detached'.[6] Due to the Queen's busy schedule and frequent absences, Charles was brought up by nannies and often in the care of his grandmother, and she might have been the one who started it all.

Most likely, Charles' first exposure to alternative medicine was in the form of the homeopathic pills he received from his grandmother when he was ill. The British royal family has, of course, a long history of believing in homeopathy. In 1832, the British physician, Frederic Hervey Foster Quin, visited Samuel Hahnemann, the father of homeopathy, in Germany. Within weeks, Dr Quin became a fully converted homeopath and returned to England with a mission. Being well-connected to aristocracy, he attracted many influential personalities to homeopathy. In 1849, a homeopathic hospital was established in in London which, in 1920, received Royal Patronage from the Duke of York, later King George VI. Later, the Queen, who had long had an official royal homeopath, was the hospital's patron, while Charles has recently become the patron of the Faculty of Homeopathy, the professional organisation of doctor homeopaths in the UK (Chapter 29).

In his book, *Harmony*[2] (Chapter 25), Charles offers further hints about his conversion to alternative medicine and his thinking behind it:

> I have tried for 25 years to encourage social and environmental responsible business; to suggest a more balanced approach to certain aspects of medicine and healthcare; more rounded ways of educating our children and a more benign, 'wholeistic' approach to science and technology... Essentially it is the spiritual dimension to our existence that has been dangerously neglected during the modern era—the dimension which is related to our feelings about things... [M]y view is that our outlook in the Westernized world has become far too firmly framed by a mechanistic approach to science... I remember that period in the 1960s only too well and even as a teenager I felt deeply disturbed by what seemed to have become a dangerously short-sighted approach. I could not help feeling that in whichever field these changes were taking hold, with industrialized techniques replacing traditional practices, something very precious was being lost... Such was the dogma of the day that when eventually, in the 1970s, I began to raise these concerns publicly, I had to face an avalanche of criticism that was nearly all based on a very basic misunderstanding. Most critics imagined that I somehow wanted to turn the clock back to some mystical Golden Age when all was a perfect rural idyll. But nothing could be further from the truth... I felt it my duty to warn of the consequences of ignoring Nature's intrinsic tendency towards harmony and balance before it was all too late... [I]f we ignore Nature, everything starts to unravel.

'One evanescent influence in the prince's life at the end of the 1970s was Zoe Sallis, a twenty year old Anglo-Italian actress...', wrote Sally Bedell Smith in her biography of Charles. 'Emotionally and spiritually attached to Sallis, Charles briefly became a vegetarian... [and he] regarded his own instincts as revealed truths superior to those of professionals and experts.'[12] It is perhaps not easy to estimate the influence of Zoë Sallis (Box 4) on Charles, but the observation that Charles, even in his late twenties, trusted his 'instincts' more than experts does certainly ring true for his support for alternative medicine in later life.

Charles' attraction to alternative medicine was clearly based more on intuition than on science. Much of it had its origin in concepts such as spiritualism and vitalism, the long obsolete metaphysical concept that life depends on a mystical vital force distinct from chemical, physical, or other principles. Vitalism is engrained in many different cultures and healing traditions, e.g. *chi* in China, *pneuma* in ancient Greece, and *prana* in India. The common denominator is the assumption that a metaphysical energy animates all living systems. Hahnemann, the originator of homeopathy, for instance, believed that homeopathic remedies act through a vital force released during the process of 'potentisation' (diluting and shaking the remedies).

Jonathan Dimbleby stated in his authorised biography of Charles[6] that 'in the latter half of the seventies, the Prince embarked on a private voyage of spiritual inquiry... An intuitive aversion to the laws of scientific materialism drew him towards the study of mysticism... he began a tentative inquiry into the field of what its practitioners referred to as "psychical research" or "parapsychology"'. Dimbleby then noted that 'in the mid-seventies, the Prince also fell under the spell of Laurens van der Post', a relationship that we will discuss in the following chapter.

BOX 4

Zoë Sallis

- Zoë Sallis was born in India and educated in England.
- She won a scholarship to the Webber Douglas Academy of Dramatic Art.
- She acted in stage performances, television, and films.
- She had a son (Danny, who today is a well-known actor) with film director John Huston.
- She has published two books, *Ten Eternal Questions* and *Our Stories, Our Visions*, which were both international bestsellers.
- Zoë introduced Charles to spiritualism and Eastern religions.

Five

Laurens van der Post

Laurens van der Post (1906–1996) was a South African writer who, as a young man, came to England, befriended the Bloomsbury group of writers, worked as a journalist, married, volunteered to serve in the British armed forces in WW2, was captured by the Japanese, and spent about three years in Japanese prisoner of war camps. After the war, He divorced his first wife and married his long-standing lover, Ingaret Giffard. She introduced him to the concepts Carl G. Jung's psychotherapy (Box 5). Ingaret and Laurens became part of Jung's circle of friends. Later, she trained as a Jungian analyst herself and even treated Princess Diana on numerous occasions.

In the 1950s, van der Post began his successful career as a writer and managed to publish several bestsellers. Charles often stated that, while still at school, he admired van der Post's works. In 1955, the BBC commissioned van der Post to go on an expedition to the Kalahari Desert in search of the elusive Bushmen, a journey that was turned into a successful six-part television series. In 1958, his perhaps best-known book was published under the same title as the BBC series: *The Lost World of the Kalahari*. Ingaret and Laurens bought a house in Aldeburgh in Suffolk, where they made friends with people who introduced them to Charles.

Laurens van der Post was a gifted storyteller, and that evidently included telling numerous and tall tales about his own life. He reinvented himself repeatedly, and his biographer, J.D.F. Jones, explains in detail how Laurens managed to spin a web of lies around virtually every aspect of his already colourful life.[13]

This fact is aptly illustrated by a short piece Laurens once wrote about himself for his publisher:

> I did not really begin to speak a civilised language until I was seven and no English until I was ten or eleven. Have spent most of my adult life with one foot in Africa and one in England. Devoted much of my moneyed leisure to getting to know Africa: there was hardly a mile of it I haven't walked, or a corner of it I have not looked into... Farmed also in England until outbreak of war when, although technically in the South African forces, I joined up as a private in England. Served in early Commandos and Special Forces...

To this, Jones comments in his authorised biography written after van der Post's death: 'His imagination was working overtime again. Not a single word of this was true.'[13] Jones also recounts van der Post's obsessive, yet unsuccessful, meddling in African politics, his plotting against Nelson Mandela, his multiple attempts to win a Nobel Prize by suggesting himself via strawmen, and his increasing inability to distinguish truth from fiction. After van der Post had died in 1996, a doctor who knew him well was asked about the cause of death. His response was illuminating: 'He had become weary of sustaining so many lies...'[13]

5.1. Charles' admiration

'It seemed to have been a union of mutual needs, between a Prince longing to find meaning in his existence and a storyteller who could weave apparent answers out of thin air.'[3] Laurens van der Post was oozing charm and charisma and sensed that 'for the Prince, there was a missing dimension', as Jonathan Dimbleby put it.[6] By 1975, the two men had formed such a close rapport that van der Post felt able to counsel him about spiritual matters, urging him to explore the 'old world of the spirit' and 'the inward way' towards truth and understanding. Van der Post suggested the two of them make a seven-week journey into the Kalahari Desert. This, he believed, would introduce Charles to the spirit world. Preparations were made in 1977 but, in the end, the plan had to be abandoned. Instead, the two later went to Kenya where they spent five days of long walks and 'intense conversation'.[3]

Van der Post urged Charles to play 'a dynamic and as yet unimagined role to suit the future shape of a fundamentally reappraised and renewed modern society', a reappraisal that would be 'so widespread and go so deep that it will involve a prolonged fight for all that is good and creative in the human imagination'.[3] An aspect of this fight, he claimed, would be 'to restore the human being to a lost natural aspect of his own spirit; to restore his relevance for life and his love of nature, and to draw closer to the original blueprint and plan of life…'[3]

Laurens left an interview for posthumous publication. In it, he expressed his hope that Charles would never become king, as this would imprison him, and it would be more important that Charles continues to be a great prince. 'He's been brought up in a terrible way… He's a natural Renaissance man, a man who believes in the wholeness and totality of life… Why should it be that if you try to contemplate your natural self that you should be thought to be peculiar?'[13]

'For 20 years they had most intimate conversations and correspondence… with a steady flow of reassurance and encouragement, political and diplomatic advice, memoranda, draft speeches and guidance for reading.'[13] Van der Post introduced Charles to the teachings of Carl Jung and his concept of the 'collective unconscious' that binds all humans together regardless whether they are Kalahari Bushmen or princes. At the behest of van der Post, Charles began to record his dreams, which van der Post then interpreted according to Jung's theories. In the late 1970s, van der Post tried to convince Charles to give up all his duties and withdraw from the world completely in search for an 'inner world truth'. This plan, too, was aborted.

All biographers agree that van der Post was the strongest intellectual influence of Charles' life.

- Charles sought van der Post's advice and spiritual guidance on numerous occasions.
- When William was born, he made van der Post his godfather.

- When Charles' marriage to Diana ran into difficulties, the couple was counselled by van der Post.
- Charles invited Laurens regularly to Highgrove, Sandringham, and Balmoral.
- Charles visited van der Post on his deathbed.
- After Laurens' death, Charles created a series of annual lectures hosted in van der Post's memory which he hosted in St James's Palace.

Laurens van der Post died two days after his 90th birthday in 1996. There had been plans for a birthday party organised by Charles. The party was cancelled at the last moment because of Laurens' illness. Today, there is a 'van der Post Memorial Garden' at his place of birth in South Africa with a section in honour of his association with the British royal family.[14]

5.2. The evidence

Some critics had long been less than enthusiastic about van der Post and his work. The *New Statesman*, for instance, reviewing one of his books, argued that 'Mr van der Post's mysticism is sometimes merely a failure to think through a genuinely important idea... It is not Jung's fault that Mr van der Post talks sheer rubbish'.[13] There were also those who warned Charles that van der Post might just be a charismatic flatterer who merely emulated what the then-Prince wanted to hear.

After his death, van der Post's reputation imploded quickly. The reasons were summarized by his biographer in *The Guardian*:

> My research showed him to be a compulsive fantasist, not just in his fiction but in the autobiographical books which he presented as non-fiction. His descriptions of his family background were fanciful; he falsely claimed that his father was a Dutch aristocrat and said his maternal grandfather had killed the last Bushman painter. (In the course of my research I discovered that, in reality, his mother was descended from a Hottentot princess, though it is unlikely that Van der Post knew this.)

He misrepresented his wartime career, claiming that he was a lieutenant-colonel when he was an acting captain. He falsely claimed that he had co-founded the Capricorn movement (a political grouping in central and east Africa which attempted to propose a multiracial solution for the region) and that he was the architect of the Rhodesian settlement in 1980. Almost all the tales Van der Post related throughout his life, and which he claimed were personally told him by a Bushman, were in fact drawn from the research of a 19th-century German scholar, Dr Wilhelm Bleek.

Time after time, the storyteller's tales about himself were inaccurate, embellished, exaggerated, distorted or invented. Put more bluntly, he was a constant liar...

Occasionally, he admitted this. In one of his last books, *A Walk With A White Bushman,* he writes: 'This is one of the problems for me: stories in a way are more completely real to me than life in the here and now. A really true story has transcendent reality for me which is greater than the reality of life. It incorporates life but it goes beyond it.'

The woman who looked after him for the last four years of his life, housekeeper Janet Campbell, later said: 'He was such an astonishing liar it seemed as automatic and necessary to him as breathing, from some flim-flam to do with socks to the engorged fabrication of his deeds. Consequently I found it impossible to see him as anything but his own invention.'

How many of these myths were also self-delusion is hard to tell, but his capacity to present a false image to others was coupled with a tendency to overestimate his own abilities...[15]

In 2001, van der Post made headlines yet again. *The Observer* reported the following:

Documents found by the biographer J D F Jones, have confirmed allegations that Van der Post had a secret child after an illicit affair with a 14-year-old girl. After the author's death in 1996, Cari Mostert sensationally came forward to claim she was his illegitimate daughter and that her underage mother had been seduced during a boat trip to England...

Mostert claimed her mother had been seduced by Van der Post, who was more than three times her age, after she had been entrusted to his care. She described the explorer's tears when they first met, and alleged he had refused to answer her letters.

The affair took place five years after Van der Post's second marriage to Ingaret Giffard in 1949, when the girl, Bonnie, travelled to London to become a ballet dancer. Only a few weeks after her arrival in England she went back, pregnant, to South Africa. Jones's research has shown the writer did, however, make financial provision for the baby, arranging to support her until she was 18 by secret deed of covenant.[16]

The story was also picked up by *The New York Times* where further details emerged:

> As for Mr. Jones's allegations about her father's relationship with a 14-year-old girl, 'I'm afraid I think that's true,' Ms. Crichton-Miller said. 'He was not a saint. He hurt people. He hurt me. But by God, he was fascinating.'
>
> Bonny Kohler-Baker, whom van der Post seduced and abandoned when she was 14, is the mother of van der Post's other daughter. She now lives outside New York City under a different name, and would not discuss the book. But her daughter, Cari Mostert, in a phone interview from the Eastern Transvaal in South Africa, said she had been brought up to believe that her maternal grandmother was her mother and that her mother was her sister. She said her grandmother had told her when she was 10 that van der Post was her father. Ms. Mostert described meeting her father for the first time when she was 12, when she and her mother had surprised him in Los Angeles, where he had a speaking engagement: 'I was crying, and he was crying.'
>
> Ms. Mostert said she had confronted him once again, as he arrived in Johannesburg airport, and he had said that her grandmother had lied in saying that she was his daughter. She claimed that she had sent her father over 50 letters, but that he had never replied. 'I thought he is such an upright, a noble

human being,' Ms. Mostert said, 'if he would only understand...' Her voice trailed off.[17]

Ever since these disclosures, Charles' silence on the subject of van der Post seems deafening. The annual lecture series in van der Post's honour was quietly cancelled. Remarkably, Charles does not mention van der Post with a single word in his book *Harmony* where, according to the publisher, it is the first time Charles' underlying philosophy has been explained.[2] The book does, however, include a section on the Bushmen of the Kalahari Desert who, according to Charles, know 'that life is a web of interconnectedness' and who 'have profound spiritual reverence for the Earth'.[2] (J.D.F. Jones assures us that 'the Prince did not meet any Bushmen'.[13])

5.3. Consequences

Charles' notions about medicine are unquestionably inspired by van der Post. Laurens, for instance, bemoaned the inadequacy of conventional medicine and wrote: 'Even if doctors did... use dreams and their decoding as an essential part of their diagnostic equipment and perhaps could confront cancer at the point of entry, how are they to turn it aside, unless they are humble enough to keep their instruments in their cases and look for some new form of navigation over an uncharted sea of the human spirit?'[13] As we will see in the next chapters, van der Post's influence shines through in many of Charles' speeches. Moreover, it contributed to the attitude of many critical observers towards Charles. Christopher Hitchens is but one example for many:

> We have known for a long time that Prince Charles' empty sails are so rigged as to be swelled by any passing waft or breeze of crankiness and cant. He fell for the fake anthropologist Laurens van der Post. He was bowled over by the charms of homeopathic medicine. He has been believably reported as saying that plants do better if you talk to them in a soothing and encouraging way... The heir to the throne seems to possess the ability to surround himself—perhaps by some

mysterious ultramagnetic force?—with every moon-faced spoon-bender, shrub-flatterer, and water-diviner within range.[18]

The following chapters will show that Hitchens might not have been far off the mark.

BOX 5

Carl Gustav Jung

- Carl Gustav Jung (1875–1961) was a Swiss analytical psychiatrist and writer.
- He had worked with and been greatly influenced by Sigmund Freud.
- However, the two later fell out over a professional disagreement.
- Jung developed a famous personality test as well as the concepts of the extraverted and the introverted personality, archetypes, and the collective unconscious.
- Jung was influenced by alchemy and the occult, claiming he could learn important things about psychology from them.
- Jung's work has been influential in psychiatry as well as in the study of religion, literature, and related fields.

Six

The British Medical Association

Shortly after his 34th birthday, in 1982, Charles was elected as President of the British Medical Association (BMA). He was therefore asked to address the festive congregation on the occasion of the 150th anniversary of the BMA's foundation. According to Jonathan Dimbleby,[6] Charles was unsure what to say at such an august gathering. 'He wandered across to his bookshelf and picked up a book about the 16th century healer Paracelsus. He read a few pages, and suddenly a host of ideas and emotions took shape...' (Paracelsus was a famous Swiss alchemist of the Renaissance period).

6.1. Charles' 1984 speech to the BMA

As the medical luminaries took their seats on that December evening, they surely expected the then-Prince to praise the many achievements of the BMA (Box 6). They were, however, in for a surprise: Charles did not bother with the customary niceties; instead, he went for an all-out frontal attack:

> I have often thought that one of the less attractive traits of various professional bodies and institutions is the deeply ingrained suspicion and outright hostility which can exist towards anything unorthodox or unconventional. I suppose it is inevitable that something which is different should arouse strong feelings on the part of the majority whose conventional

> wisdom is to be challenged or, in a more social sense, whose way of life and customs are being insulted by something rather alien. I suppose too that human nature is such that we are frequently prevented from seeing that what is today's unorthodoxy is probably going to be tomorrow's convention. Perhaps we just have to accept that it is God's will that the unorthodox individual is doomed to years of frustration, ridicule and failure in order to act out his role in the scheme of things, until his day arrives and mankind is ready to receive his message, which he probably finds hard to explain himself, but which he knows comes from a far deeper source than conscious thought...[19]

Almost as though van der Post (Chapter 5) was dictating his words, Charles went on to state that there was 'a message for our time, a time in which science has become estranged from Nature', and he continued:

> [The Doctor] should be intimate with Nature. He must have the intuition which is necessary to understand the patient, his body, his disease. He must have the 'feel' and the 'touch' which makes it possible for him to be in sympathetic communication with the patient's spirits... Through the centuries healing has been practised by folk healers who are guided by traditional wisdom which sees illness as a disorder of the whole person, involving not only the patient's body, but his mind, his self-image, his dependence on the physical and social environment, as well as his relation to the cosmos... The good doctor's therapeutic success largely depends on his ability to inspire the patient with confidence and to mobilise his will to health... the whole imposing edifice of modern medicine, for all its breath-taking success, is, like the celebrated Tower of Pisa, out of balance...

6.2. Reactions

One does not have to be in the grips of a particularly vivid fantasy to imagine how, listening to Charles, the assembled luminaries must have felt: 'his words sent a shudder through the medical establishment', writes Dimbleby.[6] Another biographer, Tom Bower, is more direct: 'For a 34 year old with a mediocre degree in history to preach the power of spiritualism, based on a jumble of ideas inherited from van der Post, Jung and the 16th century Swiss healer and alchemist Paracelsus, and then to announce, to an audience of doctors, the science that produced the drugs that had eradicated polio and tuberculosis, was as frightening as it was brazen.'[20]

Charles himself, and some sympathetic commentators, often implied that the speech was a much-needed wake-up call for the fossilised medical profession that set things in motion, started a discussion, and soon opened the door to more acceptance of alternative medicine. Yet, there is good evidence to argue that it achieved the exact opposite. It did not soften the fronts, but it hardened them. It did not build a bridge, but it endangered the bonds that already existed. Getting lectured like first-year medical students by a young man who evidently was ill-informed can hardly have amused the seasoned physicians.

By 1982, the medical profession had, of course, long recognised the importance of 'the "feel" and the "touch" which makes it possible for him to be in sympathetic communication with the patient's spirits'. The profession had also considered the 'unorthodox or unconventional' seriously:

- Psychosomatic medicine had become an established part of healthcare.
- In 1982, Medline, the largest medical database, listed thousands of articles on all sorts of alternative medicine.
- Britain had five homeopathic NHS hospitals where various forms of alternative medicine were practised.
- Balint had published his seminal work *The Doctor, His Patient and the Illness* some 30 years earlier.
- Balint groups for doctors were all the rage in the medical profession.
- Medical schools ran courses for future doctors focused on the therapeutic relationship.

- The proverbially abrupt, non-communicative surgeon had become a caricature of days gone by.

Charles' outburst therefore risked a counterproductive step backwards reinforcing barriers that had all but vanished. The affront prompted a reluctance of the UK medical establishment to look benevolently at alternative medicine—much in contrast to Germany, for instance, where numerous initiatives were emerging at medical schools achieving the exact opposite.

In her biography, Catherine Mayer points out that Charles' speech also contained a reasonable passage: 'Wonderful as many of them are, it should still be more widely stressed by doctors that the health of human beings is so often determined by their behaviour, their food and the nature of their environment.' Mayer claims that 'this was not only sensible stuff; it needed—and still needs—to be said'.[3] She may well have a point! There is doubtlessly much to be criticised about modern medicine's over-emphasis on drug treatment. But was the public celebration of the BMA's 150th anniversary truly the right occasion to voice criticism? Had Charles used his new position as the president of this powerful organisation wisely, he could have, for instance, initiated a working group to look into the complex issue of over-prescribing. In this way, he might even have prevented or minimised our current problems with opiate addiction.

Was Charles' speech merely a lapsus of an inexperienced youngster? Perhaps, but this pattern of behaviour would henceforth repeat itself with some regularity. In pursuing an aim, Charles tends to impatiently assail experts rather than drawing them on his side for productive cooperation. What seems to be missing is an insight into his own limitations and lack of expertise as well as the tolerance of allowing experts to censure his ideas. It appears that as soon as one of his advisors deviates from Charles' own creed, he or she is replaced by someone who is more compliant. The sad result of this process is that, over the years, Charles has surrounded himself with a circle of seemingly docile yes-men who are unable or unwilling correct even his most obvious errors. As mentioned in Chapter 4, Charles 'regarded his own instincts as revealed truths superior to those of professionals and experts'.[12]

Charles' lack of insight into his own limitations is aptly demonstrated by the way he reflected on his BMA speech almost two decades later. In 2001 he reminisced:

> I remember being asked to be president of the BMA in their 150th anniversary year in 1982, going to the dinner and making a speech... all I did was to beg for a little bit more understanding about the need for a more holistic approach towards the way in which we carry out our healthcare. I quoted a certain amount from Paracelsus and at the end of it all I was absolutely astonished to find what a reaction it had caused amongst the medical establishment. All hell broke loose![21]

6.3. Consequences

The reaction of the BMA to Charles' affront in 1982 was all too predictable: the doctors felt challenged, perhaps even insulted by someone who had used the festive occasion for displaying his own ignorance of their work. All hell did evidently not break loose; instead, the BMA politely tried to make the best of the difficult situation and established a working group charged with looking at alternative medicine and producing a full report on it.

The eight-expert-strong panel of the 'BMA Board of Science and Education' did not take its task lightly; they called for information and evidence from doctors and alternative therapists. Over 600 submissions were received and distilled into their report which considered the main alternative therapies in detail. The final report spoke of 'a revision to primitive beliefs and outmoded practices, almost all without basis', and concluded that 'alternative therapies [have] little in common between them except that they pay little regard to scientific principles of orthodox medicine and indeed may contravene them'.[22]

It took several years until another attempt to assess alternative medicine was made by the BMA. In 1993, the BMA Scientific Division conducted a second review of alternative medicine and published another report on the subject.[23] Based on evidence that had emerged in the meanwhile, it turned out to be much more welcoming to alternative medicine and its practitioners. It included statements such as:

- Closer collaboration between the medical profession and practitioners of non-conventional medicine in clinical research should be encouraged.
- The BMA recommends that a single regulatory body be established for each therapy.
- The BMA recommends that the General Medical Council be asked to consider whether doctors be permitted to refer patients for specific treatments to registered practitioners.
- The BMA in particular recommends that accredited post-graduate sessions be set up to inform clinicians on the techniques used by different therapists and the possible benefits for patients.

This progress, however, was made not because of but despite Charles' ill-considered intervention in 1982.

BOX 6

The British Medical Association (BMA)

- The BMA's stated aim is 'to promote the medical and allied sciences, and to maintain the honour and interests of the medical profession'.
- It has about 159,000 doctors and 19,000 medical students as members.
- The BMA represents, supports, and negotiates on behalf of all UK doctors and medical students.
- It is a member-run and led organisation that campaigns on the issues impacting the medical profession.
- The organisation traces its origins to the Provincial Medical and Surgical Association, founded by Sir Charles Hastings in 1832. It became the BMA in 1836.
- *The Provincial Medical and Surgical Journal* was started in 1842 and became the *British Medical Journal* (BMJ) in 1857.
- Today, the BMJ is one of the most respected general medical journals in the world.

Seven

Talking Health

Between April 1984 and January 1987, a total of eight colloquia took place at the Royal Society of Medicine (RSM) in London, a 'leading provider of high-quality continuing postgraduate education and learning to the medical profession and wider healthcare teams'.[24] The meetings were chaired by Sir James Watt, the past president of the RSM and a former surgeon general of the British Navy.[25] The colloquia were entitled 'Talking Health' and focused entirely on alternative medicine. Together with experts from the RSM and other members of the medical profession, UK representatives of six alternative therapies took part:

- osteopathy,
- chiropractic,
- acupuncture,
- homeopathy,
- naturopathy,
- medical herbalism.

Sir James's introduction started with the following words: 'In his address to the British Medical Association in 1982, the Prince of Wales voiced his fear that our current preoccupation with the sophistication of modern medicine would divert our attention from "those ancient, unconscious forces lying beneath the surface, which still help to shape the psychological attitudes of modern man".' The series of colloquia was thus a direct reaction to Charles' divisive BMA speech of 1982 (Chapter 6). In more than one way, it seemed to be the attempt of several well-meaning UK members of the medical establishment to smooth over the upset it had caused.

Charles attended some of the meetings in person and also contributed to the discussions. He was then invited to write the foreword of the book summarising the proceedings which was published by the RSM in 1988.[26]

7.1. Charles' foreword

The short text of Charles' foreword is remarkable, predominantly for its tone. It implies that he regretted the hoo-ha he had caused by his 1982 speech at the BMA and thus tried to hit a more conciliatory note:

> Having provoked a certain amount of discussion during the 150th anniversary of the British Medical Association, I was most encouraged when the Royal Society of Medicine decided to hold a series of Colloquia to debate what I consider to be a most important aspect of the total health care of individuals.
>
> The Royal Society of Medicine has therefore rendered a valuable service to the future development of health care. Through hosting these Colloquia, your Society had given an opportunity, probably for the first time, for practitioners, both conventional and complementary, to come together to learn how each other think and work.
>
> The importance of communication has been emphasised by all practitioners who took part and it is particularly pleasing to observe how effective these Colloquia have been. Difficulties and differences are inevitable when dialogue of this nature is undertaken. The Colloquia on research and training are, I am sure, the preliminary exchanges in a much longer debate.
>
> Scientific progress comes as much through deductive logic, rational debate and critical evaluation as it does through intuitive reasoning, creative play and the ability to tolerate uncertainty. The reports on the discussion indicate that both these essential strands of progress were maintained.
>
> It has given me great pleasure to take part, albeit peripherally, in these Colloquia and I very much hope that, as

a result of these gatherings, better ways will be found to deal with the myriad of problems that patients have. I also hope that, in the long run, a situation will develop by which the general practitioner can recommend genuine complementary therapists to treat his patients, if he feels that this is the more appropriate thing to do. I need hardly to say that I look forward to the continuation of the work that is so well describes in this volume.

7.2. The evidence

Compared to his speech of 1982 at the BMA (Chapter 6), Charles' text was subdued. It did mention his preoccupation with intuitive reasoning, of course (Box 7). But the accusing tone, the aggressive challenge to conventional medicine, the questioning of reductionist science had all but disappeared.

The published proceedings of the Colloquia contain many further conciliatory and, to my mind, remarkable statements, for example:

- 'Both conventional and complementary practitioners agree that the needs of the patient are paramount...'
- '...all good practitioners approach their patient "holistically"...'
- 'The medical profession does not view complementary medicine with any real hostility...'
- '[There is a] need for controlled clinical trials to validate complementary therapies...'
- 'We must look for a single standard of high quality research...'
- 'Medicine is not exclusively science. The art of medicine has long been acknowledged...'
- '...training in complementary medicine should be comparable to that in orthodox medicine...'
- '...we should attempt to overcome methodological shortcomings by improved scientific skills, for the alternative would be to fall back into an age of dogmatic pseudoscience...'

- '...research [is] a crucial way of validating complementary therapies and of protecting the public...'

The 'Talking Health' Colloquia were thus a constructive attempt to rekindle the dialogue between the UK medical orthodoxy and alternative medicine which Charles had previously disrupted. There was plenty of time and opportunity for all sides to be heard.

- Did Charles learn a lesson?
- Was the moderation of his tone merely temporary?
- Had he become more realistic in his ambitions?

The next chapters will go some way towards answering these questions.

Re-reading the proceedings today leaves me impressed with the number of issues that were addressed already all those years ago. Unfortunately, it also demonstrates how very few of the problems debated in 1988 have been resolved today. This, I feel, highlights in an exemplary fashion the futility of the endless debates we conducted during the 34 years that followed and the irrationality of relentlessly promoting types of healthcare that lack sound evidence. What has been so often missing in such debates is a clear consensus what the priorities are. It evidently fascinates some people interested in alternative medicine to discuss at length:

- the popularity of alternative medicine,
- the reasons for its popularity,
- the sociological context in which alternative medicine exists,
- the acceptance of alternative medicine by doctors,
- the motivations of patients to try alternative medicine,
- etc., etc.

All of these themes are no doubt interesting but, without sound evidence to prove that alternative medicine generates more good than harm, these discussions all too often become irrelevant and are little more than a waste of time.

7.3. Consequences

The Colloquia were one of the very rare occasions where Charles listened to and discussed with experts who did not necessarily share his views on alternative medicine. I know of no other event where this happened since.

Within the field of alternative medicine in the UK, the meetings were considered a resounding success. They started the dialogue between two camps which continues to the present day. They provided an occasion for both sides to realise that the 'opposition' is not nearly as unapproachable or hostile as they may have previously thought. And the realists of both camps might have understood that progress can only occur after reliable evidence demonstrating that alternative medicine does more good than harm has emerged.

For me personally, the Colloquia turned out to be a life-changing event, even though I was not present at the time. During one of the discussions, the philanthropist Sir Maurice Laing apparently commented that, in his view, UK alternative medicine would not go anywhere unless a proper professorial chair was created in one of the British universities entirely dedicated to this subject. Impressed with this idea, Sir James later managed to persuade Sir Maurice to endow such a post. The University of Exeter then accepted the project by creating the first Chair of Complementary Medicine on the planet. Thus, my Department of Complementary Medicine was born in 1993. Sir James Watt became one of its official advisors and, despite his advancing years, served in this function (visiting us regularly in Exeter) until shortly before his death.

Box 7

Intuitive reasoning

- Intuition is defined as the ability to understand or know something based on feelings rather than fact.
- Intuitive reasoning is more than using common sense.
- It is often referred to as gut feeling, sixth sense, inner sense, instinct, inner voice, spiritual guide, etc.
- Intuitive reasoning is fundamentally different from rational reasoning; the latter is based on evidence, the former on emotions and feelings.
- Intuitive reasoning is not a valid basis for arriving at generalisable decisions about the validity of medical interventions.
- Many things in medicine are overtly counter-intuitive.

Eight

Osteopathy

Osteopathy is a manual treatment invented over a century ago by the American healer Andrew Taylor Still (1828-1917). Today, US osteopaths (doctors of osteopathy or DOs) have more or less stopped practising manual therapy and are fully recognised as medical doctors who can specialise in any medical field after their basic training, which is similar to that of MDs. In the UK and most countries, however, osteopaths still practice almost exclusively manual treatments according to Still's instructions and are thus considered alternative practitioners. Whenever Charles talks about osteopathy, he refers to the latter category.

Andrew Still defined osteopathy diffusely as

> ...a science which consists of such exact, exhaustive, and verifiable knowledge of the structure and function of the human mechanism, anatomical, physiological and psychological, including the chemistry and physics of its known elements, as has made discoverable certain organic laws and remedial resources, within the body itself, by which nature under the scientific treatment peculiar to osteopathic practice, apart from all ordinary methods of extraneous, artificial, or medicinal stimulation, and in harmonious accord with its own mechanical principles, molecular activities, and metabolic processes, may recover from displacements, disorganizations, derangements, and consequent disease, and regained its normal equilibrium of form and function in health and strength.[27]

More than 100 years later, osteopathy is still not defined more concisely. The General Osteopathy Council, the governing body of UK osteopaths, offers an equally vague definition:

> Osteopathy is a system of diagnosis and treatment for a wide range of medical conditions. It works with the structure and function of the body, and is based on the principle that the well-being of an individual depends on the skeleton, muscles, ligaments and connective tissues functioning smoothly together.
>
> To an osteopath, for your body to work well, its structure must also work well. So osteopaths work to restore your body to a state of balance, where possible without the use of drugs or surgery. Osteopaths use touch, physical manipulation, stretching and massage to increase the mobility of joints, to relieve muscle tension, to enhance the blood and nerve supply to tissues, and to help your body's own healing mechanisms. They may also provide advice on posture and exercise to aid recovery, promote health and prevent symptoms recurring.[28]

In fact, osteopathy is not dissimilar to chiropractic (Chapter 9). It relies heavily on manual manipulation and mobilisation of the spine. One important difference, however, is that osteopaths tend to use less frequently those manual techniques that are associated with serious adverse effects. Osteopathy therefore tends to be considerably less dangerous than chiropractic.[29]

The first person to introduce osteopathy to the UK was John Martin Littlejohn, through his annual lectures in London from 1898 to 1900. Later, American-trained osteopathic practitioners came to work in England, Scotland, and Ireland. In 1913, Littlejohn returned to live in England and in 1917 he opened the British School of Osteopathy in London, the first osteopathic school in Europe.[30]

8.1. Charles' support of osteopathy

Not least through the efforts of Charles, UK osteopaths became in 1993 the first profession of alternative practitioners to be regulated by statute. At the time, this development amazed many healthcare professionals; osteopathy was neither well supported by evidence (see below), nor were there many osteopaths practising in the UK. Charles later commented: 'osteopathy... [is] now regulated in the same way as doctors and dentists, with their own Acts of Parliament. I'm very proud to have played a tiny role in trying to push for that Act of Parliament over the years.'[31]

The links of UK osteopathy to the royal family go back many years: the British School of Osteopathy, recently renamed the University College of Osteopathy, had royal support from the outset. King Charles is the patron of the UK General Osteopathic Council, and Princess Anne is the patron of the British School of Osteopathy.

Next to homeopathy (Chapter 29), osteopathy seems to have a special place in Charles' heart; in his book *Harmony* (Chapter 25), these are the two alternative therapies he discusses in most detail. About osteopathy, he states that 'it is a fascinating method of treatment'.[2] Andrew Still 'was empathic that disease was a disturbance in the natural flow of blood; the nerves cause muscles to contract, which compresses the blood returning to the heart', Charles explains. 'Osteopaths have a skilled, specialised sense of touch that can detect an altered quality of motion in the tissues of the body. Recognizing the state of disharmony in the body, they work to restore balance towards a state of harmony and health. In this way, osteopathy offers an integrated and preventive system of healthcare. It understands the unique connection between the physical, mental and spiritual make-up of each person.'[2]

8.2. The evidence

The quote above highlights Charles' less than factual view of human physiology. And why not, some people might ask. It is, of course, not essential that the heir to the throne understands how our bodies function. Neither is it necessary that he comprehends the need for evidence before calling a therapy a 'preventive system of healthcare'. Yet, the general public who he addresses with such statements might be less confused if he (or his advisors) would make sure that his pronouncements are factually correct.

The truth is that the basic assumptions that underpin osteopathy are not biologically plausible and many of the claims made by osteopaths are steeped in pseudoscience. An overview of 100 randomly selected websites of osteopaths, for example, revealed that 93% of them fulfilled at least one of the criteria for pseudoscientific claims.[32] Some osteopaths consider themselves to be back pain specialists, while others treat a much wider range of conditions. In either case, the evidence is unconvincing or outright negative:

- For back pain, the evidence is encouraging but not conclusively positive. One review (by osteopaths) concluded that osteopathic treatment 'significantly reduces low back pain. The level of pain reduction is greater than expected from placebo effects alone and persists for at least three months. Additional research is warranted to elucidate mechanistically how osteopathic manipulative exerts its effects, to determine if osteopathic manipulative treatment benefits are long lasting, and to assess the cost-effectiveness of osteopathic manipulative treatment as a complementary treatment for low back pain'.[33]
- An independent and more critical review, however, found that the data fail to produce compelling evidence for the effectiveness of osteopathy as a treatment of musculoskeletal pain.[34]
- A systematic review of osteopathy for spinal conditions included 19 randomised trials. The authors concluded that '[t]oday, no clear conclusions of the impact of osteopathic care for spinal complaints can be drawn'.[35]
- For non-spinal conditions, the evidence is even less convincing. One review concluded, for instance, that the evidence of the effectiveness of osteopathic manipulative therapy for paediatric conditions remains unproven due to the paucity and low methodological quality of the primary studies.[36]
- Specifically for disease prevention, there is no evidence at all to support Charles' above-quoted statement.

Even some leading osteopaths acknowledge this lack of sound evidence. When, in 2017, the UK School of Osteopathy was given university status, its principal was quoted saying: 'We recognise that for some of the things that some osteopaths are doing, there is very limited evidence [to demonstrate their effectiveness]...'[37] And even NICE no longer recommends osteopathy as a first line treatment for back pain.[38]

Osteopathy also is not free of side effects. Severe complications including cauda equina syndrome, lumbar disk herniation, fracture, and haematoma or haemorrhagic cyst have been noted after osteopathic manipulations. Contraindications include conditions with increased risks of bleeding or diseases that compromise bone, tendon, ligament, or joint integrity.[39] Remarkably, osteopaths have no monitoring system in place that would allow us to define how often adverse effects occur.

8.3. Consequences

Has osteopathy started to prosper as a result of Charles' lobbying? Have osteopaths made meaningful contributions to public health? Has the NHS adopted osteopathy wholesale? Is osteopathy recommended by NICE for treating back pain or any other condition? The uniform answer to these questions is no.

Perhaps the simplest and most transparent measure to judge the success of osteopathy in the UK is to look at the amount of research published by UK osteopaths each year. It ranged from one (!) in 1988 to its current maximum of five (!) in 2016, and the University College of Osteopathy seems to have an average output of published papers of around three per year.[40]

Yes, Charles' support has brought the statutory regulation of osteopathy in the UK, but it has led to precious little else. And one thing is worth remembering: even the best regulation of nonsense must result in nonsense.

Box 8

Andrew Taylor Still

- Andrew Taylor Still (1828–1917) is the founder of osteopathy.
- During the American Civil War, Still served as a hospital steward.
- In his autobiography, Still stated that he served as a 'de facto surgeon', but there appears to be no record of him obtaining a medical degree.
- Still realised that the medical practices of his day often caused significant harm and studied to find a new way of healing.
- He thus dabbled in alternative treatments, such as hydropathy, diet, magnetic healing, and bone-setting.
- The latter approach, he perfected into osteopathy.
- By diagnosing and treating the musculoskeletal system, Still believed he could treat a wide range of conditions.
- In 1892, Still founded the first school of osteopathy, the American School of Osteopathy (now called A.T. Still University) in Kirksville, Missouri.

Nine
Chiropractic

According to its originator, David Daniel Palmer, chiropractic is 'the art of adjusting by hand all subluxations of the three hundred articulations of the human skeletal frame, more especially the 52 articulations of the spinal column, for the purpose of freeing impinged nerves, as they emanate thru the intervertebral foramina, causing abnormal function, in excess or not, named disease'.[41]

Palmer believed that practically all diseases are caused by subluxations of the spine and thus can only be cured by spinal 'adjustments', i.e. manual manipulations of the spine (Box 9). Today, the chiropractic profession is divided into those who believe the gospel of their founding father and those who see themselves are pure back pain specialists.

9.1. Royal support for chiropractic

Royal support for chiropractic has been evident for some time. In 1990, HRH Diana, Princess of Wales, became the Patron of the Anglo-European College of Chiropractic.[42] During this period, Lord Kindersley arranged influential dinners with Charles, who wanted to bring about the statutory regulation of osteopathy (Chapter 8) and chiropractic in the UK. Jonathan Dimbleby wrote: 'With the help of Lord Kindersley, he [Charles] arranged a private dinner... for leading figures... Amongst whom were a chiropractor, a herbalist and the presidents of all the leading medical colleges...'[6] Ian Hutchinson, former Chair of the Chiropractic Steering Group, Member of the King's Fund Working Party on Chiropractic, and former President of the British Chiropractic

Association, commented: 'In 1983/84 I met two people who were to help us over the years, Lord Kindersley and Sir James Watt, at dinner parties held by HRH The Prince of Wales.'[43]

A Working Party on Chiropractic eventually recommended statutory regulation, and the presentation at the Kings Fund was attended by Charles.[44] In 1994, UK chiropractors thus became regulated by statute and gained a legal status equal to doctors, dentists, nurses, or physiotherapists. The main driver behind this unusual (and in the eyes of most independent experts undeserved) upgrade was Charles' lobbying.[45] Charles confirmed:

> As we know, the professions of Osteopathy and Chiropractice are now regulated in the same way as doctors and dentists, with their own Acts of Parliament. I'm very proud to have played a tiny role in trying to push for that Act of Parliament over the years. It has also been reassuring to see the progress being made by the other main complementary professions and I look forward to the further development of regulatory frameworks enabling high standards of training, clinical practice and professional behaviour.[31]

Charles' original plan was to also obtain statutory regulation for homeopaths and herbalists.[46] This, however, never happened.[47]

In 2013, 'The Royal College of Chiropractors' (RCC) was created.[48] A royal charter is a formal document issued by the monarch, granting a right or power to an individual or a corporate body. It is used for establishing significant organisations such as cities or universities. The RCC describes its own history as follows:

> The legislation underpinning the UK chiropractic profession, the Chiropractors Act, received Royal assent on 5th July 1994 and the Privy Council announced the membership of the General Chiropractic Council (GCC), the profession's registering body, on 28th January 1997.
>
> During the intervening period, it became clear that the Act would not fulfil all the aspirations of the profession in terms of moving into the mainstream of healthcare; there was an

obvious gap between the GCC as the registering body and the professional associations acting as trade unions. This gap related to such areas as postgraduate education and training, research and specialisation.

On the advice of a senior medical figure, an organisational model similar to that of a Medical Royal College was devised. Thus, the College of Chiropractors was conceived during 1997 and incorporated in 1998 as an independent body to develop, encourage and maintain the highest possible standards of chiropractic practice for the benefit of patients.

Over the next couple of years the embryonic 'College' grew with a regional faculty infrastructure, the mainstay of the organisation, becoming firmly established in order to foster education locally. As an independent body, separate from any of the political groups, members were able to share information and expertise from all areas of the profession. Following its incorporation in October 1998, the College of Chiropractors was formally launched on 28th April 1999 at the King's Fund…

At a meeting of the Privy Council on Wednesday 12th November 2012, the Queen approved the grant of a Royal Charter to the College, the first Royal Charter to be granted to a complementary medicine organisation in the UK…

9.2. The evidence

Vertebral subluxations (malalignments of spinal structures which, according to chiropractic belief, require spinal manipulation) are the cornerstone of chiropractic 'philosophy' and the *raison d'être* of chiropractors. Yet, today there is a broad consensus that subluxations, as understood by chiropractors, do not even exist.[49] Chiropractors who nevertheless believe in subluxations diagnose them in nearly 100% of the population—even in individuals who are completely free of spinal abnormalities and symptoms. Consequently, almost all patients consulting a chiropractor receive spinal manipulations. In addition, most chiropractors also employ a range of other non-drug treatments mostly borrowed from physiotherapy.

Today, some chiropractors focus on treating back and neck pain. So-called straight chiropractors, however, claim to be primary care physicians who can treat almost any condition. Yet, the evidence that chiropractic spinal manipulation is effective beyond placebo is weak for spinal problems such as back and neck pain and negative for all other conditions.[50] Most chiropractors also recommend maintenance care, i.e. regular spinal manipulations, even in the absence of symptoms. There is, however, no good evidence to show that maintenance care is effective (except for increasing the cash flow of chiropractors).[51]

Chiropractic spinal manipulations cause mild to moderate adverse effects, such as pain, in about 50% of all patients.[52] In addition, it is associated with very serious complications, usually caused by neck manipulation damaging an artery that supplies parts of the brain, resulting in a stroke and even death. Several hundred such cases have been documented in the medical literature.[52] As there is no system in place to monitor such events, the true figure is almost certainly much larger.[41] To these direct risks, we have to add important indirect risks of chiropractic; they relate to the fact that some chiropractors tend to:

- fail to appreciate the limits of their competence by treating patients (e.g. infants) and conditions they know little about;
- issue anti-vaccination or other harmful advice and misinformation;
- advise patients against taking prescribed drugs;
- disregard the contra-indications of spinal manipulations (e.g. osteoporosis);
- violate medical ethics by not obtaining fully informed consent from their patients;
- overuse X-ray diagnostics;
- treat patients longer than necessary to boost their income;
- sell useless dietary supplements.

Of course, such concerns apply to some other alternative healthcare professions as well. They are listed here because there is good evidence to suggest that, in the realm of chiropractic, they are particularly relevant.[41]

The following statement of the RCC is an example of chiropractors being in denial about their potential to do harm:

> Experiencing mild or moderate adverse effects after manual therapy, such as soreness or stiffness, is relatively common, affecting up to 50% of patients. However, such 'benign effects' are a normal outcome and are not unique to chiropractic care.
>
> Cases of serious adverse events, including spinal or neurological problems and strokes caused by damage to arteries in the neck, have been associated with spinal manipulation. Such events are rare with estimates ranging from 1 per 2 million manipulations to 13 per 10,000 patients...[53]

The statement is misleading in several respects. Experiencing mild or moderate adverse effects after chiropractic spinal manipulations, such as pain or stiffness usually lasting 1–3 days, does impair patients' quality of life.[52] To call such problems benign and normal is unjustified and irresponsible. Serious adverse events, including spinal or neurological problems and strokes, happen at an unknown rate. As there are no surveillance systems, nobody can tell how often they occur. An association between stroke caused by vertebral artery damage or 'dissection' (VAD) and chiropractic spinal manipulation has been reported in about 20 independent investigations. Many people have lost their lives at the hands of chiropractors. The deaths of the US model, Katie May, and the UK pensioner, John Lawler, are just two recent and widely-reported examples.[54,55]

In view of this evidence, one must question whether chiropractic generates more good than harm. Charles' unreserved, long-term support for chiropractic is therefore surprising to say the least.

9.3. Consequences

As the question whether chiropractic generates more good than harm is unanswered, one would expect that Charles' support provided the much-needed boost to research aimed at answering it. However, the research activity of UK chiropractors is minute and has remained low during the years following statutory regulation in 1994. In 2020, for instance, the number of published papers authored by UK chiropractors amounted to a total of just 34. In fact, chiropractors seem to view their elevation to statutory regulation as a substitute for evidence. When Simon Singh exposed the fact that chiropractors 'happily make bogus claims', the British Chiropractic Association, instead of rectifying this situation by producing solid evidence, sued him for libel.[56]

One might also have expected that Charles' interventions would have prompted UK chiropractors to set up a monitoring system for adverse effects of spinal manipulations. Despite the risks of chiropractic manipulations being known for decades, no such system has yet been established.

Finally, one would have hoped that statutory regulation led to an adequate governance of UK chiropractors. Sadly, this does not seem to be the case. When the Professional Standards Authority audited the GCC in 2014, the verdict was disappointing:

> The extent of the deficiencies we found in this audit (as set out in detail above) which related to failures across every aspect of the casework framework, as well as widespread failures to comply with the GCC's own procedures, raises concern about the extent to which the public can have confidence in the GCC's operation of its initial stages FTP process.[57]

It seems, therefore, that Charles' support of chiropractic has brought about few if any positive developments and no tangible benefits to public health.

Box 9

The colourful life of David Daniel Palmer (1845–1913)

- 1867: DD Palmer starts as a teacher in Concord, Iowa.
- 1869: DD and his younger brother TJ start their career as beekeepers in Letts, Iowa.
- 1871: DD marries Abba Lord who calls herself a 'psychometrist, clairvoyant physician, soul reader and business medium' (she leaves him in 1873).
- 1876: DD marries Louvenia Landers, a widow; they have 4 children together, including BJ who later becomes DD's partner in his chiropractic business (she dies in 1884).
- 1880: DD publishes a pamphlet about spiritualism and refers to himself as a 'spiritualist'.

Chapter 9

- 1885, 25 May: DD marries Martha Henning. The marriage is short-lived; on 8 July of the same year, DD posted a public notice in the 'What Cheer Patriot' disowning her.
- 1886: DD moves to Iola, Kansas, where he practices as a magnetic healer and calls himself 'Dr Palmer, healer'.
- 1886: DD advertises his services as a 'vitalist healer'.
- 1887, 25 October: one of DD's patients has died and there is an inquest.
- 1887: DD moves to Davenport and advertises: 'DD Palmer, cures without medicine…'
- 1888, 6 November: DD marries Villa; they stay together until her death in 1905.
- 1894: DD publishes his views on smallpox vaccination: '…the monstrous delusion… fastened on us by the medical profession, enforced by the state boards, and supported by the mass of unthinking people …'
- 1895: DD starts a business selling gold fish.
- 1895, 18 September: DD administers the first spinal manipulation to Harvey Lillard; this day is celebrated as the birth of chiropractic.
- 1898: DD opens his first school of chiropractic in Davenport, the 'Palmer School of Chiropractic', which has survived to the present day.
- 1902, 27 April: DD first used the term 'subluxation' in a letter to his son BJ ('…where you find the greatest heat, there you will find the subluxation causing the inflammation which produces the fever…').
- 1902, 6 September: DD is arrested in Pasadena when a patient suffering from consumption dies after DD's second adjustment; the charges were later dropped because of a technicality.
- 1903 DD is charged with practising medicine without licence but, before the case goes to trial, DD leaves for Chicago.
- 1904, December: DD starts his journal, *The Chiropractor*, which survives until 1961. DD's very first article is entitled '17 Years of Practice'.
- 1905: DD's former students Langworthy and Smith accuse DD of stealing the concepts of chiropractic from the Bohemian bonesetters of Iowa.
- 1905: DD's wife Villa overdoses on morphine and dies; the coroner is unable to tell whether she committed suicide or intended it for pain relief.
- 1906: DD marries Mary Hunter.
- 1906: DD is again on trial for practising medicine without a licence and found guilty. The penalty is US$350 or 105 days in jail. DD chooses jail. However, his new wife, Mary, bails him out after 23 days.
- 1906: BJ and DD publish their book, *Science of Chiropractic*; DD claims that most of the chapters were written by him.
- 1907: DD opens another grocery store after falling out with BJ.
- 1908: DD opens the 'Palmer College of Chiropractic' in Portland, Oregon.
- 1910: DD publishes his book *The Chiropractor's Adjuster*.
- 1911: DD toys with the idea of turning chiropractic into a religion.
- 1913: DD visits Davenport where he is injured. Mary later accuses BJ of striking his father with his car and thus causing his death.
- 1913, 20 October: DD dies; the official cause of death is typhoid fever, a condition that he repeatedly claimed to be curable by a single spinal adjustment.
- 1914: DD Palmer's book *The Chiropractor* is published.

Ten
The Foundation of Integrated Health

In 1993, Charles founded an organisation which, after being renamed several times, ended up being called the 'Foundation for Integrated Health' (FIH).[58]

10.1. Charles' intentions

Charles established the charity to explore 'how safe, proven complementary therapies can work in conjunction with mainstream medicine'.[59] This might sound like the remit of a research institute, however, as it turned out, research was not of much interest to Charles' foundation (and investigating the safety of alternative therapies even less). Wikipedia is much more accurate: 'The Foundation promoted complementary and alternative medicine, preferring to use the term "integrated health", and lobbied for its inclusion in the National Health Service.'[58]

In 1999, Charles spoke at a conference organised by his foundation and elaborated: 'I look back to the rather "lukewarm" response I received in 1983 as President of the British Medical Association when I first spoke about integration and complementary and alternative medicine. We have clearly travelled a very long way since that time. I believed then, as I do now, that the move to a more integrated provision of healthcare would ultimately benefit patients and their families.'[60] In 2000, he was similarly upbeat, stating that 'The Foundation for Integrated Medicine is establishing an information data base and resource centre for practitioners and professionals who want to develop integrated services. It is no longer a question of hacking a path through unexplored jungle. The way is landmarked, the goals and prizes are visible. All that is needed is courage—and a little imagination.'[61]

In yet another one of his speeches, Charles said: 'I actually believe that the integration of the best of ancient and modern approaches can be of benefit in the relief of unnecessary suffering on the part of patients.'[62]

10.2. The evidence

As we will discuss in more detail in Chapter 13, there are several significant problems with the notion of integrating the 'best of ancient and modern approaches':

- It is based on belief, not science.
- Without evidence, 'the best' is an arbitrary choice and a meaningless term.
- The evidence that alternative medicine relieves suffering is less than solid or, in many cases, even negative.

During its 17 years of existence, the FIH organised numerous meetings, published various documents, lobbied to increase the use of alternative medicine within the UK National Health Service (NHS), and was involved in drawing up statutory regulation for acupuncturists and herbalists (which never materialised).[63] To finance this work, the FIH received several large grants from the UK government, and Charles regularly motivated wealthy businessmen to donate additional money. Funds were also accepted from several manufacturers of alternative medicine products. Between 2005 and 2007, the charity's annual turnover was about £1.2 million.[64] The funds were spent mostly on activities that one could loosely summarise as lobbying for alternative medicine. Contrary to the impression that Charles liked to give, the FIH did not sponsor meaningful amounts of research into alternative medicine.

One achievement of the FIH was the formation of the 'Complementary and Natural Healthcare Council' (CNHC) which aimed to register 10,000 practitioners by the end of 2009. By September 2009, it had, however, succeeded in enrolling less than a tenth of that number.

On numerous occasions, the FIH divided the scientific and medical community over its campaigns promoting alternative medicine—for instance, by encouraging general practitioners to offer herbal and other alternative treatments to their patients, or by publishing guides for the general public which were misleading consumers by not being objective and complete about the evidence. At the time, I commented on the FIH's patient guide that '[t]he guide is largely an evidence-free space. It certainly does not inform patients about issues such as effectiveness and potential harms of treatments. Instead, it is full of promotional and pseudoscientific language'.[65]

While I got involved in several FIH activities (for instance, I invited them regularly to our annual research conference at Exeter), our relationship was often strained; they aimed at promoting the use of alternative medicine, while I felt that such endeavours were irresponsible before scientific tests had proven that alternative medicine generated more good than harm.

In July 2003, I saw an announcement published in the foundation's newsletter:

> The Peninsula Medical School aims to become the UK's first medical school to include integrated medicine at postgraduate level. The school also plans to extend the current range and depth of programmes offered by including healthcare ethics and legislation. Professor John Tooke, dean of the Peninsula Medical School, said: 'The inclusion of integrated medicine is a patient driven development. Increasingly the public is turning to the medical profession for information about complementary medicines. This programme will play an important role in developing critical understanding of a wide range of therapies'.

This was the first I knew about a course offered by my medical school on my subject. I was then told that it had been initiated at the behest of Charles and was being organised by Dr Michael

Dixon[66] (Box 10). The funding came from a manufacturer of homeopathic remedies. I refused to get involved with the course which, only months after it had been created, closed.[5]

On 19 March 2010, I was quoted stating that, 'in my view, the field of complementary medicine would benefit greatly, if the FIH ceased to exist and the funds thus freed were directed towards rigorous research'.[5] Only weeks later, accounting irregularities were noted by the foundation's auditors and the Metropolitan Police Economic and Specialist Crime Command began an inquiry into alleged fraud and money laundering.[67] Two officials at the FIH were arrested and the foundation announced that it would close:

> 30 April 2010
> The Trustees of The Prince's Foundation for Integrated Health have decided to close the charity. Whilst the closure has been planned for many months and is part of an agreed strategy, the Trustees have brought forward the closure timetable as a result of a fraud investigation at the charity.
>
> The Trustees feel that The Foundation has achieved its key objective of promoting the use of integrated health. Since The Foundation was set up in 1993, integrated health has become part of the mainstream healthcare agenda, with over half a million patients using complementary therapies each year, alongside conventional medicine.
>
> From 2000–2007, at the request of the Department of Health, The Foundation ran a regulation programme which resulted in the creation, in 2008, of an independent self-regulatory body for complementary therapy, called the Complementary and Natural Healthcare Council.
>
> On 1st April 2010, the Secretary of State for Health announced plans to introduce statutory regulation for herbalists and to consider the equivalent for acupuncture.
>
> The Trustees believe that the best way of promoting integrated healthcare in the future is through the networks of specialist practitioners which the charity has helped to establish.

These networks have brought together specialists and proponents of integrated healthcare, such as doctors, nurses, clinicians, consultants, scientists and students.

The announcement was skilfully spun; yet it is hard to deny that the closure was forced by the discovery of false accounting.[68]

10.3. Consequences

The Foundation's former finance director, George Gray, was sentenced to a three-year custodial sentence for siphoning off £253,000 of FIH's funds.[3] Substantial amounts of tax payers' money got wasted on meaningless projects. Charles' reputation received a major hit. Of all the defeats Charles had to endure while lobbying for alternative medicine, the closure of the FIH must be amongst the most significant.

After the FIH's closure, I was invited to write a short epitaph for *The Independent*:

> The Foundation for Integrated Health is closing. Should we be sad? I don't think so. During recent years, this organisation seems to have broken charity regulations by becoming a tool for Prince Charles to implement his often bizarre ideas regarding healthcare.
>
> He seems to think that the nation should be force-fed on alternative medicine today, while research into these treatments might be conducted some time in the future. I, on the other hand, have often pointed out that research has to come first; it should sort out the wheat from the chaff and, subsequently, we might consider integrating those treatments that demonstrably generate more good than harm. I therefore think that the FIH has become a lobby group for unproven and disproven treatments populated by sycophants.
>
> The FIH has repeatedly been economical with the truth. For instance when it published a DoH-sponsored patient guide that was devoid of evidence. They claimed evidence was never meant to be included. But I had seen a draft where it

was and friends have seen the contract with the DoH where 'evidence' was an important element.

I hope that, after the demise of FIH, the discussion about alternative medicine in the UK can once more become rational. I also hope that Prince Charles has the greatness of selecting advisers who actually advise rather than 'Yes Men' who are hoping to see their names on the next Honours List.[69]

After the foundation had been liquidated, its former medical director, Michael Dixon, established the 'College of Medicine', later to be renamed 'College of Medicine and Integrated Health' (Chapter 23). Charles was initially not formally associated with the new organisation. However, in 2019, he became its patron.

Box 10

Dr Michael Dixon LVO, OBE, MA, FRCGP

- Studied medicine at Guy's Hospital.
- Senior partner in a GP practice in Devon.
- Chair of the NHS Alliance (1998–2015).
- Medical director of the FIH at the time when it was shut down.
- Chair of the College of Medicine and Integrated Health.
- Advisor to Charles.
- Head of the Royal Medical Household.
- Author of several books. Charles wrote the foreword to one stating that 'for Dr Michael Dixon, the mission of promoting health has been a life-long passion'.[70]
- Author of a study of spiritual healing concluding that 'healing may be an effective adjunct for the treatment of chronically ill patients presenting in general practice'.

Eleven

Open Letter to The Times

In 2000, Charles published an open letter in *The Times* entitled 'When Our Health is at Risk, Why Be So Mean?'[71] It was one of the first times that he elaborated to the general public on his views specifically about research into alternative medicine. The text reveals Charles' beliefs in an exemplary fashion. I therefore reproduce it here in full. I have merely added 15 references (in square brackets) which refer to my explanations and comments below.

11.1. The letter

Complementary medicine has come a long way since the British Medical Journal dismissed it as a 'flight from sciences' and suggested chiropractic was no better than the 'examination of a bird's entrails'.

As many as one in five of us opts at some time for such treatments, and the relationship between complementary therapies and science is distinctly warmer. But the thaw is not without lumps of ice. A report by the Lords Select Committee on Science and Technology has highlighted an urgent need for more research into alternatives to orthodox medical practice.[1]

It makes good sense to evaluate complementary and alternative therapies. For one thing, since an estimated £1.6 billion is spent each year on them[2], then we want value for our money. The very popularity of the non-conventional approaches suggests that people are either dissatisfied with their orthodox treatment, or they find genuine relief in such therapies.[3] Whatever the case, if they are proved to work, they should be made more widely available on the NHS.[4]

There is already a greater acceptance of complementary medicine among healthcare professionals. About 40 per cent of GP practices offer access to some form of non-conventional treatment such as osteopathy or homoeopathy.[5]

Acupuncture is increasingly routine in pain and rheumatology clinics, and in more than 90 per cent of hospices soothing therapies such as massage and aromatherapy are available. NHS cancer patients at Charing Cross and Hammersmith Hospitals can receive a wide choice of treatments from the complementary therapies team—reflexology, aromatherapy, massage therapy, relaxation training and art therapy.[6]

But there remains the cry from the medical establishment of 'where's the proof?'—and clinical trials of the calibre that science demands cost money. For instance, there should be sufficient numbers of subjects—a trial that involves 300 or 3,000 people will obviously carry more weight than one with 13 or 30.[7] And there should be a control, against which the treatment being tested is shown to perform better than the current treatment of choice.

The truth is that funding in the UK for research into complementary medicine is pitiful. NHS primary care groups and health authorities are reluctant to spend large sums on non-conventional approaches without evidence of cost-effectiveness and efficacy. But because so few complementary and alternative therapies are available on the NHS, there is little incentive to divert scarce funds into research. Truly a Catch-22 situation.[8]

Much of the research funding for conventional medicine is from eminent bodies that are usually reluctant to muddy their reputation by delving into unorthodox waters, and from rich pharmaceutical companies. But few non-conventional therapies involve medication, and even where they do—as with herbal and homoeopathic remedies — the manufacturers are small companies with limited funds for research.[9]

So where can funding come from? Many of the serious studies into complementary and alternative medicine have been carried out abroad. In the United States, for example, the Government's National Centre for Complementary and Alternative Medicine (NCCAM) has a research budget of $68 million (£42.5 million) a year, expected to rise to more than $78 million in 2001. The money is used to fund 11 research centres across the country which evaluate alternative treatment for chronic health conditions such as asthma, arthritis and addictions.[10] NCCAM also collaborates with other government organisations; for example, with the US National Cancer Institutes.

And in the UK? Figures from the department of complementary medicine at the University of Exeter show that less than 8p out of every £100 of NHS funds for medical research was spent on complementary medicine.[11] In 1998–99 the Medical Research Council spent no money on it at all, and in 1999 only 0.05 per cent of the total research budget of UK medical charities went to this area.

The Arthritis Research Campaign is one of few such organisations to take account of the huge rise in the numbers of people using non-conventional therapies. It has announced funding into complementary and alternative therapies, beginning with a two-year clinical trial into the effects of acupuncture on patients with osteoarthritis of the knee.

What, then, is to be done? A national strategy for complementary and alternative medicine research would be a start.[12] With new funding, the Foundation for Integrated Medicine, of which I am the president and founder, could provide a focus to co-ordinate this strategy, allocate funding, provide a networking resource, train researchers, disseminate information and monitor research development.[13]

But serious funding is also needed for bursaries, fellowships, research centres linked to higher education institutes and to support 'fledgeling' researchers, whether complementary practitioners with little experience of trial protocols, or old science hands unschooled in complementary medicine.

At the same time, we should be mindful that clinically controlled trials alone are not the only prerequisites to apply a healthcare intervention. Consumer-based surveys can explore why people choose complementary and alternative medicine and tease out the therapeutic powers of belief and trust.[14]

A potentially powerful resource is at our fingertips, but its benefits will be limited—and often those who can least afford to pay for complementary medicine are the ones who would most benefit[15]—unless somewhere, somehow, purses are opened and funds dedicated to its systematic study.

11.2. My explanations and comments

1. The report on alternative medicine by the House of Lords (Box 11) highlights the fact that the evidence for alternative therapies is mostly negative and does not suggest they generate more good than harm.[72]
2. The figure is from research published by my Exeter team.[73] This proves that Charles and his advisors did follow our work. Yet, Charles never cited our papers if they contradicted his beliefs or purpose.
3. The research on this subject does, in fact, not suggest these factors to be the most important reasons for consumers to try alternative medicine: 'Along with being more educated and reporting poorer health status, the majority of alternative medicine users appear to be doing so not so much as a result of being dissatisfied with conventional medicine but largely because they find these health care alternatives to be more congruent with their own values, beliefs, and philosophical orientations toward health and life.'[74] Similarly, the report by the House of Lords stated that 'the holistic approach of CAM, the individual emphasis, the greater time spent on patients by practitioners... [are] all very popular

reasons... for patients' satisfaction with complementary and alternative medicine'.[72]
4. The crucial point here is that the Lords' report failed to demonstrate that they 'are proved to work'.[72]
5. The acceptance of alternative medicine has remained fairly constant and did, in fact, not significantly increase over time.[75]
6. It is wrong to imply that reflexology is an effective treatment; it is based on irrational assumptions and lacks proof of effectiveness for any condition.[76]
7. This is not necessarily true. The optimal sample size for a clinical trial is best determined via a calculation based on data from a pilot study.
8. Research funds could, and perhaps should, also come from those who earn their living with alternative medicine; for instance, via contributions from professional organisations of alternative practitioners.
9. Some companies have an annual turnover of tens, even hundreds, of millions.
10. Even today, the NCCAM has not come up with a single alternative medicine that is demonstrably effective.[77]
11. Another result of our research[78] – see point 2 above.
12. Another one of Charles' ideas that never materialised.
13. Even though there had been ample demand and opportunity, Charles' foundation never did get involved in funding research.
14. The field of alternative medicine has long suffered from too many (mostly useless) surveys.[79]
15. There is no good evidence to suggest that either the rich or the poor would significantly benefit from using more alternative medicine.

11.3. Consequences

Charles revealed himself as being very selective in his choice of evidence. He called for more research and research funding, without actually sponsoring research, or understanding that research is supposed to test hypotheses (rather than confirm his beliefs), or (worst of all) acting according to the findings of existing research. Charles' article disclosed once again that he had long made up his mind: alternative medicine is effective whatever research might show. And therefore, presumably, his foundation was never tasked to sponsor research; its focus was on promoting alternative medicine usage. Charles' call for more research therefore turns out to be a red herring and little more than a fig leaf to cover his own deficits in rational thinking.

Crucially, we need to ask whether Charles' call to fund research in alternative medicine had any major effect? Have substantial amounts of money been made available as a result? The answer is a simple 'No'.

Box 11

Important quotes from the House of Lords' report regarding evidence on alternative medicines

- 'Complementary/alternative medicine (CAM) has been criticised... for not having scientific evidence to back its claims.'
- '...[L]ittle research is being done, and... the few studies which have been completed are given disproportionate weight.'
- 'More research is needed on the efficacy of most CAMs.'
- '...[A]ny discipline whose practitioners make specific claims for being able to treat specific conditions should have evidence of being able to do this above and beyond the placebo effect. This is especially true for therapies which aim to be available on the NHS and aim to operate as an alternative to conventional medicine...'
- '...[T]he risk/benefit ratio of the therapy in question must be considered. If the potential benefits of a therapy are likely to be very significant, or even life-saving, then the level of risk a patient may be willing to take with the therapy is likely to be higher than the level of risk they are willing to accept for the benefit of temporary symptom relief or the cure of a minor complaint.'
- '...[P]atient satisfaction... alone cannot be taken as a proof or otherwise of a treatment's efficacy or as evidence to justify provision.'

Twelve

The Model Hospital

Charles had plans for a model hospital of integrated medicine. In the early 2000s, he had been discussing with Dr Mosaraf Ali the possibility of giving official royal support to such a venture.[80] Eleanor Stoikov, Dr Ali's clinical manager, commented; 'The prince is giving support, but not in a financial way.' And a spokesperson for St James's Palace said: 'It [the hospital for integrative medicine] is an interesting proposition and they have had private discussions on the matter. The prince has argued for some time for a greater role for integrated medicine.'[81]

12.1. Charles' concept

The 'new age hospital' was claimed to be historically unprecedented and unique. It was meant as a place for training doctors to combine conventional medicine with alternative practices, such as homeopathy, Ayurvedic medicine, and acupuncture. The new hospital was to have around 100 beds. The then-prince's intervention marked the culmination of years of campaigning by him for the NHS to assign a greater role to alternative medicine. The new hospital was planned to be overseen by Dr Mosaraf Ali.

12.2. The evidence

Historically, such a concept had been tried many years before. During the Third Reich, a very similar project had been realised — the Rudolf Hess Krankenhaus in Dresden (1934–1941). Under the banner of the 'Neue Deutsche Heilkunde' (Box 12), it combined conventional healthcare with homeopathy and various other types of alternative medicine, favoured by several of the Nazi elite (not least Rudolf Hess,[82] Hitler's deputy at the time). The hospital did not, however, survive for long; one of the reasons for its failure may well have been the lack of sound evidence supporting the use of alternative medicine.[83] The historical precedent is nevertheless interesting, as it was the first large-scale attempt to create the type of integrated medicine that Charles promotes today.[84]

Charles' new hospital was due to open in London in 2003/04 under the direction of Mosaraf Ali, a man with an interesting background. Mosaraf Ali was born in India, where he also started his medical studies. In 1973 Ali got a scholarship to study medicine in the former USSR. He then joined the Central Institute of Advanced Medical Studies in Moscow for postgraduate study in acupuncture. During that period, he specialised in pulse diagnosis, iridology, tongue diagnosis, and hypnosis. He returned to Delhi in 1982 to study Ayurveda, Unani medicine, yoga, and marma therapy (Box 27). In 1991, Ali came to the UK and took up a post at the Hale Clinic in London, and in 1998 he opened his own Integrated Medical Centre in London. Ali boasts that his 'unique knowledge and experience makes him one the leaders in his field of medical practice… These qualities made him a Royal Physician and family doctor of many highly influential public figures worldwide'.[85] He also claims that with one look he has the power to heal with no medicine and that he diagnoses with methods that include looking into the eyes (iridology, Chapter 25).[86] Ali uses the title 'doctor' but the General Medical Council, which registers doctors in the UK, claims he isn't registered with them.[87]

Ali likes to recount the moment he first met Charles:

> I met the Prince by chance at a reception. At that time, I was working at the Hale Clinic in London. He was very interested in my work and soon afterwards, I began to treat the entire royal family. For the past 14 years, I have been a member of the medical team of the British royal family. The family strongly believes in alternative medicine. They exercise, eat a healthy diet and spend a great deal of time outdoors in nature.

This is why they are so healthy and long-lived. I often say that my treatment methods are not medicine at all, but a way to help keep people healthy.[88]

'Dr. Ali has done us a great favour in pointing out the way forward during the coming centuries and I hope his books will help to revolutionize many people's lives and reintegrate mind, body and the spirit', Charles wrote in the foreword of Ali's 2001 book modestly entitled *The Integrated Health Bible*.[89] 'I am wholly convinced that his approach has the power to really make an immense difference to many people's lives and to our society as a whole', Charles added.

In the end, the project of the new model hospital never materialised. Might this be due to Mosaraf Ali falling into disrepute?[90] In 2005, a 69-year-old stroke patient, Raj Bathija, consulted Dr Mosaraf Ali. According to documents lodged with the High Court, Mr Bathija was given 'marma massage' (Box 27) and told to eat potassium-rich foods. (At the time no studies of marma massage existed. Charles later initiated a pilot study of marma massage at Exeter.[5] Its results failed to show that the therapy is effective for stroke patients.[91]) Ali's patient claimed that, after this treatment, he was left in pain and with his right foot pale and cold.[92] Four days after receiving the marma massage, Mr Bathija was admitted to hospital where both of his legs were amputated due to severe leg ischaemia.[93] According to the patient, Ali and his brother were negligent in that they failed to diagnose his condition and neglected to advise him to go to hospital. Bathija's daughter Shibani said: 'My father was in a wheelchair but was making progress with his walking. He hoped he might become a bit more independent. With the amputations, that's all gone.'[86]

12.3. Consequences

Charles' model hospital is yet another project in the realm of alternative medicine that failed to be the huge success envisioned. In fact, it never materialised. Dr Ali continues to practice but Charles hardly ever mentioned him after the court case.

Box 12

'Neue Deutsche Heilkunde' (New German Medicine)

- In the 1930s, Germany had as many non-medically trained alternative practitioners as it had doctors.
- Several of the 'Nazi elite', e.g. Hess and Himmler, were ardent proponents of alternative medicine.
- They tried to unify German medicine under the banner of 'Neue Deutsche Heilkunde', which essentially was an integration of conventional and alternative medicine.
- In 1938, the Nazis created the 'Heilpraktiker', a healthcare profession of medically untrained alternative practitioners which still exists today.
- The model hospital of the 'Neue Deutsche Heilkunde' was the 'Rudolf Hess Krankenhaus' in Dresden.
- The Nazis also initiated a huge research project to validate homoeopathy. Its results were devastatingly negative.[84]

Thirteen

Integrated Medicine

In 2001, Charles published his first short editorial in the *British Medical Journal* introducing his ideas around integrated medicine. Its title: 'The Best of Both Worlds'. In it he stated that 'The concept of integration... starts with the integration within each individual of the body, mind, and spirit through to the holistic development of the whole person... Integrated medicine is more than simply about curing disease and symptoms. It is about encouraging individual responsibility for one's own health'.[94] Ever since, Charles has been internationally recognised as one of the world's most prominent champions of integrated medicine (or integrative medicine, as it is called in the US).

13.1. Charles' views on integrated medicine

In a 2012 editorial, Charles updated his views on integrated medicine and expressed them more fully.

> ...By integrated medicine, I mean the kind of care that integrates the best of new technology and current knowledge with ancient wisdom. More specifically, perhaps, it is an approach to care of the patient which includes mind, body and spirit and which maximizes the potential of conventional, lifestyle and complementary approaches in the process of healing...
>
> ...I have been attempting to suggest that it might be beneficial to develop truly integrated systems of providing health and care. That is, not simply to treat the symptoms of disease, but actively to create health and to put the patient at

the heart of this process by incorporating those core human elements of mind, body and spirit...

This whole area of work—what I can only describe as an 'integrated approach' in the UK, or 'integrative' in the USA—takes what we know about appropriate conventional, lifestyle and complementary approaches and applies them to patients. I cannot help feeling that we need to be prepared to offer the patient the **'best of all worlds'** according to a patient's wishes, beliefs and needs. This requires modern science to understand, value and use patient perspective and belief rather than seeking to exclude them—something which, in the view of many professionals in the field, occurs too often and too readily...

Now, surely, is the time for us all to concentrate some real effort in these areas. We will need to do so by deploying approaches which, at their heart, retain the crucial bedrock elements of traditional and modern civilized health care—of empathy, compassion and the enduring values of the caring professions.[95]

In 2021, Charles updated his views yet again and wrote:

For as long as I care to remember, I have suggested that medicine should become more integrated and inclusive... Many patients choose to see complementary practitioners for interventions such as manipulation, acupuncture and massage. Surely in an era of personalised medicine, we need to be open-minded about the choices that patients make and embrace them where they clearly improve their ability to care for themselves?... I have always advocated 'the best of both worlds', bringing evidence-informed conventional and complementary medicine together and avoiding that gulf between them, which leads, I understand, to a substantial proportion of patients feeling that they cannot discuss complementary medicine with their doctors...[96]

13.2. The evidence

As pointed out in Chapter 12, Charles' concept of combining conventional with alternative medicine is less original or progressive than one might think. In 1937, Rudolf Hess, Hitler's deputy, opened the World Congress of Homeopathy in Berlin making similar points (my translation):

> I ask the medical profession to consider previously excluded therapies with an open mind. It is necessary that an unbiased evaluation takes place, not just of the theories but also of the clinical effectiveness of alternative medicine... Insightful doctors, some of whom famous, have, during the recent years, spoken openly about the crisis in medicine and the dead end that health care has manoeuvred itself into. It seems obvious that the solution is going in directions which embrace nature. Hardly any other form of science is so tightly bound to nature as is the science occupied with healing living creatures. The demand for holism is getting stronger and stronger, a general demand which has already been fruitful on the political level. For medicine, the challenge is to treat more than previously by influencing the whole organism when we aim to heal a diseased organ.[97]

According to Charles and other enthusiasts, integrated medicine is based on two main concepts. The first is that of 'whole person care', and the second is often called 'the best of both worlds'. Attractive concepts, some might think, but do these two pillars of integrated medicine stand up to scrutiny?

Whole patient care

Practitioners of integrated medicine do not just treat the physical complaints of a patient but they claim to look after the whole individual: body, mind, and soul. On the surface, this holistic approach seems most laudable and attractive. Yet a closer look reveals major problems.

The truth is that all good medicine is, was, and always will be holistic (Charles was told this already during the 'Talking Health' colloquia in the mid-1980s,[26] but he must have forgotten—Chapter 7). Today's GPs, for instance, should care for their patients as whole individuals and address not just physical problems but also social and spiritual issues. Many doctors might neglect the holistic aspect of care; and some even lack compassion, empathy, and kindness. If that is so, they are, by definition, not good doctors. And, if such deficits are recognised to be widespread, we must reform conventional healthcare. Delegating holism, compassion, empathy, and kindness to integrated medicine practitioners cannot be the solution; it would mean abandoning an essential element of good healthcare and would turn out to be a serious disservice to today's patients and to the healthcare of tomorrow.

It follows that the promotion of integrated medicine under the banner of holism, compassion, etc. makes no sense at all. Either it misleads patients into believing that these features are the exclusive characteristics of integrated medicine, while, in fact, they are hallmarks of any good healthcare. Or, if holism, compassion, etc. are neglected in a particular branch of conventional medicine, it distracts us from the important task of correcting these deficits.

The best of both worlds

The second concept of integrated medicine was described by Charles as 'the best of both worlds'. Proponents of integrated medicine do indeed claim to use the 'best' of the world of alternative medicine and combine it with the 'best' of conventional healthcare. Again, this concept looks commendable at first glance but, on closer inspection, serious doubts emerge.

Those doubts hinge on the definition of the term 'best'. We have to ask, what does 'best' mean in the context of healthcare? Surely it cannot mean the most popular or fashionable—and certainly 'best' is not determined by decree of the heir to the throne. Best can only signify 'the most effective' or, more precisely, 'being associated with the most convincingly positive risk/benefit balance'.

If we understand 'the best of both worlds' in this way, it turns out to be synonymous with the concept of evidence-based medicine (EBM), the currently accepted thinking in healthcare (Box 13). And if 'the best of both worlds' is synonymous with EBM, we clearly don't need this duplicity of concepts in the first place; it would only confuse patients and distract healthcare professionals from the auspicious efforts of continuously improving healthcare. In other words, the second axiom of integrated medicine is as nonsensical as the first.

The practice of integrated medicine

So, on the basis of these somewhat theoretical considerations, integrated medicine is but a superfluous, misleading, and counterproductive distraction. The most powerful argument against integrated medicine is, however, not a theoretical but a practical one, namely the nonsensical and often dangerous treatments that are being used in its name day in, day out.

If we look around us, go on the internet, read the relevant literature, or walk into an integrated medicine clinic in our neighbourhood, we inevitably find that behind the politically correct slogans of holism and 'best of both worlds' there lurks what can only be described as pure quackery. If you don't believe me, please go and look for yourself. I promise you will discover the promotion of any unproven and disproven therapy that one can think of, anything from crystal healing to Reiki, and from homeopathy to urine-therapy. And not only that; you might even see how integrated medicine practitioners actively undermine public health, for instance by advising against vaccinations.[98] When I googled 'integrated medicine clinic, UK' in April 2021, for example, the first establishment that came up was a London-based clinic offering, amongst other treatments, homeopathy, detox, and mistletoe therapy for cancer, none of which are demonstrably effective and all of which can cause considerable harm.[99]

13.3. Consequences

Despite its lugubrious precedent during the Third Reich, the mantra of integrated medicine has today been enthusiastically embraced by most sections of alternative medicine. Yet, outside the realm of alternative medicine, it caused confusion and prompted criticism.

The confusion relates firstly to the terminology: is it integrated or integrative medicine? Which of the many definitions of integrated medicine that have been put forward[100] do actually apply, and do they make sense? What about integrated medicine as it is traditionally used to signify the integration of health and social services?

The criticism relates firstly to the fact that integrated medicine tries to hijack concepts of conventional medicine and distracts from reforming healthcare by delegating essential concepts to practitioners of integrated medicine. Secondly, it is about the practice of integrated medicine which has been aptly summarised as follows: 'If you integrate quackery with real medicine, you do not produce better medicine. Instead, you turn quackery into medicine and medicine into quackery.'[101]

Any critical review of the data will conclude that there is precious little evidence to suggest that integrated medicine has generated meaningful benefits, and a sizeable amount of evidence shows that it is has become a 'paradise for charlatans',[102] which all too often even endangers public health.[98] After careful consideration of all these arguments, I conclude that integrated medicine is one of the most colossal deceptions of healthcare today.[103]

Box 13

Evidence-based medicine

- Evidence is the body of facts that leads to a given conclusion.
- Evidence-based medicine is defined as the integration of best research evidence with clinical expertise and patient values.
- It thus rests on three pillars: external evidence, the clinician's experience, and the patient's preferences.
- External evidence must be reliable, and there is a well-established hierarchy that goes from simple opinion, to non-experimental observations, to clinical trials, to systematic reviews of the totality of the clinical trial data.
- Charles' views on alternative medicine are almost exclusively based on opinion, the lowest level in this hierarchy.

Fourteen

The Gerson Therapy

The Gerson therapy has long been promoted as an 'alternative cancer cure'. It consists of ingesting raw and organically-grown vegetables and freshly-pressed vegetable juices (up to 13 large glasses per day). In addition, regular coffee enemas are administered, allegedly stimulating the liver to detoxify the body. Finally, a range of supplements is added to the mix. The treatment is administered at high costs in specialised hospitals.

The therapy was developed about 100 years ago by German doctor Max Gerson (Box 14), who assumed that the therapy, originally developed as a treatment for tuberculosis, had cured his migraines. Subsequently he promoted it for all sorts of illnesses, even cancer. 'The Gerson Institute' claim that '[o]ver the past 60 years, thousands of people have used the Gerson Therapy to recover from a variety of illnesses, including: Cancer (including melanoma, breast cancer, prostate cancer, colon cancer, lymphoma and more), Diabetes, Heart disease, Arthritis, Autoimmune disorders, and many others'.[104]

14.1. Charles' promotion of the approach

Charles has promoted the Gerson therapy on several occasions. 'We must push Gerson!', he allegedly exclaimed over cups of tea to Kim Lavely, the chief executive of his Foundation for Integrated Health.[20] In 2004, while addressing conventional doctors at the Royal College of Gynaecology in London, Charles warmly recommended the Gerson diet to the baffled physicians as a treatment for cancer. Here are his exact words:

I know of one patient who turned to Gerson therapy having been told she was suffering from terminal cancer and would not survive another course of chemotherapy. Happily, seven years later, she is alive and well. So it is vital that, rather than dismissing such experiences, we should further investigate the beneficial nature of these treatments.[62]

14.2. The evidence

Professor Baum, an eminent oncologist, who was present during the lecture (as was I), responded to Charles' words in an open letter invited by the *British Medical Journal*:

> ...Over the past 20 years I have treated thousands of patients with cancer and lost some dear friends and relatives to this dreaded disease... The power of my authority comes with knowledge built on 40 years of study and 25 years of active involvement in cancer research. Your power and authority rest on an accident of birth. I don't begrudge you that authority but I do beg you to exercise your power with extreme caution when advising patients with life-threatening diseases to embrace unproven therapies...[105]

Baum was rightly irritated by Charles' promotion of the Gerson therapy. There was (and still is) no good evidence that the Gerson therapy is effective for cancer or any other condition. A recent review stated that no conclusions about the effectiveness of the Gerson therapy, either as an adjuvant to other cancer therapies or as a stand-alone treatment, can be drawn from any of the existing studies. The only clinical trial that has been published suggested not a prolonged but a reduced survival time for patients following this therapy.[106]

The Gerson diet is essentially a starvation diet. It deprives cancer patients of vital nutrients; in addition, it drastically impairs their quality of life. Coffee enemas remove potassium from the body and have caused serious problems:

- infections
- dehydration

- fits
- salt and other mineral imbalances in the body
- heart and lung problems, even death
- constipation and inflammation of the bowel (colitis) from regular, long-term use of enemas which can weaken the bowel muscle
- loss of appetite
- diarrhoea
- abdominal cramps
- aching, fever, and sweating
- cold sores
- dizziness and weakness

Most patients find the diet exceedingly hard to follow, and those who fail to adhere to it are often told that it is their fault if their cancer does not respond. Thus, they die prematurely not merely deprived of their funds and quality of life but also feeling guilty.

The treatment's assumed mode of action is far from plausible. Cancer Research UK sums up the situation succinctly: 'There is no scientific evidence that Gerson therapy can treat cancer. In fact, in certain situations Gerson therapy can be very harmful to your health. The diet should not be used instead of conventional cancer treatment.'[107]

14.3. Consequences

Charles' 2004 speech did not make him friends in the oncology community; they unanimously sided with Professor Baum. Today, the Gerson diet has remained on the far-out fringes of alternative medicine. Its proponents had 100 years to come up with convincing evidence that it is effective. They have failed dismally and no responsible cancer organisation in the world would recommend this therapy.

Not least thanks to Charles' support, the Gerson therapy has numerous over-enthusiastic followers who are convinced of its effectiveness and recommend it to cancer patients. The number of lives this has cost could be tragically high.

Box 14

Max Gerson (1881–1959)

- Born to a Jewish family in Wongrowitz, German Empire (now in Poland).
- Medical degree from Freiburg, Germany.
- Practised as an Internist in Breslau and Bielefeld.
- Specialised in the dietary treatment of tuberculosis.
- In 1927, he began claiming that his diet cured cancer.
- In 1933, he moved first to Vienna, then to Paris, London, and finally to New York.
- Gerson believed that diseases are caused by the accumulation of toxins in the body.
- His treatment was aimed at eliminating these toxins.
- In 1958, Gerson published a book in which he describes 50 patients who were allegedly cured of cancer.[108]
- All independent attempts to verify these outcomes have failed.
- The Gerson treatment has caused several deaths.
- The Gerson clinics on the American continent are located across the US border in Mexico, because they are illegal in the US.

Fifteen

Herbal Medicine

Herbal medicines are those with active ingredients made from plant parts, such as leaves, roots, or flowers. But being 'natural' doesn't necessarily mean they're safe for you to take. Just like conventional medicines, herbal medicines will have an effect on the body, and can be potentially harmful if not used correctly. They should therefore be used with the same care and respect as conventional medicines.[109] This description from the NHS is, of course, correct but it ignores an often overlooked yet important fact: there are two fundamentally different kinds of herbal medicine.[110]

- The first type essentially employs well-tested herbal remedies against specific conditions; this approach has been called 'rational phytotherapy' by some experts.[111] An example is the use of St John's Wort for depression, which Charles mentions regularly, for instance, in his speech for the WHO (Chapter 17). Further examples are listed in Box 15.
- The second type of herbal medicine is often called 'traditional herbal medicine'. It entails consulting a herbal practitioner who takes a history, arrives at a diagnosis (usually according to obsolete criteria and concepts), and concocts a mixture of several herbal remedies that are tailor-made to the characteristics of his patient. Thus 10 patients with an identical diagnosis, say depression, might receive 10 different mixtures of herbs, none of which might contain St John's Wort, the only evidence-based herbal treatment for this condition. Individualised

herbalism exists in several traditions, e.g. Chinese, Japanese, Indian, or European, and virtually every traditional herbalist would employ this individualised approach.

As we will discuss below, the difference between the two types of herbalism is stark: one is supported by some good evidence, while the other is not.

15.1. Charles' protection of herbal medicine

As long as Charles has been promoting alternative medicine, he has been advocating herbalism; and he never differentiates between the two types. In 2000, he made a particularly poignant pro-herbalism gesture by planting an acre of land with medicinal plants on his estate in Highgrove.[21]

After Charles had successfully lobbied for statutory regulation of both osteopathy and chiropractic in the early 1990s (Chapters 8 and 9), he tried to achieve the same for UK herbalists. In 2005, Charles wrote to the prime minister, Tony Blair, about the European Union's Directive on Herbal Medicines.[112,113,114]

> 24 February 2005
> Tony Blair
>
> Dear Prime Minister,
>
> We briefly mentioned the European Union Directive on Herbal Medicines, which is having such a deleterious effect on complementary medicine sector in this country by effectively outlawing the use of certain herbal extracts. I think we both agreed this was using a sledgehammer to crack a nut. You rightly asked me what could be done about it and I am asking the Chief Executive of my Foundation for Integrated Health to provide a more detailed briefing which I hope to be able to send shortly so that your advisers can look at it. Meanwhile, I have given Martin Hurst a note suggesting someone he could talk to who runs the Herbal Practitioner's Association.
>
> Yours ever, Charles

The Prime Minister replied politely promising to comply with Charles' wishes:

> 30 March 2005
> Response from Tony Blair
>
> Dear Prince Charles
>
> Thanks too for your contacts on herbal medicines who have been sensible and constructive. They feel that the directive itself is sound and the UK regulators excellent, but are absolutely correct in saying that the implementation as it is currently planned is crazy. We can do quite a lot here: we will delay implementation for all existing products to 2011; we will take more of the implementation upon ourselves; and I think we can sort out the problems in the technical committee — where my European experts have some very good ideas. We will be consulting with your contacts and others on the best way to do this, we simply cannot have burdensome regulation here.
>
> Yours ever, Tony

In the same year, Dr Michael Dixon, then medical director of the FIH and Charles' advisor on alternative medicine, warned of the EU restrictions on herbal remedies: 'we fear that we will see a black market in herbal products.'[115] In 2009, Charles held talks with the UK Health Secretary, asking him to introduce safeguards amid a crackdown by the EU that would have prevented anyone who is not a registered health practitioner from selling herbal remedies.[116] Charles and other campaigners seemed to forget that the EU was merely trying to protect consumers from the dangers of toxic herbal remedies.

15.2. The evidence

'Rational phytotherapy' is supported by some reasonably good evidence. This evidence is, however, limited to about one dozen of the many thousands of herbal medicines available today (Box

15).[117] For individualised or traditional herbalism, the situation is dramatically different: there is no sound evidence that this approach generates more good than harm. Some herbalists claim that individualised herbalism cannot be tested in clinical trials. This notion can easily be shown to be wrong by pointing out that several studies testing individualised herbalism have been published. Their results fail to show that individualised herbalism of any tradition is effective for any condition.[118,119]

Not only has the approach of traditional herbalists of using mixtures of multiple individualised herbal ingredients not been shown to be effective, it also carries considerably more risks than the use of a single herb. Because individualised herbalism employs a multitude of ingredients, the likelihood of adverse effects and herb–drug interactions,[120] contamination, adulteration,[121] etc. is significantly increased.

As to the regulation of herbal medicine, we should remember a simple rule applies: even the best regulation of nonsense must result in nonsense. Professor David Colquhoun, FRS, rightly commented that Charles' suggestions were wrong and would not safeguard patients: 'We do need regulation, but not the sort of ineffective regulation the Prince wants. Proper regulation should be on whether these products work. The rules should be the same as for all drugs.'[122]

15.3. Consequences

Herbal medicine is one of the few areas within the realm of alternative medicine that is, at least to some extent, supported by sound evidence. In some countries, for instance Germany, it is therefore thriving, and herbal medicines are frequently prescribed by doctors and other healthcare professionals. In the UK, this is not the case.

For once, there would have been an alternative treatment that Charles could have supported, based on sound evidence. It would have required a clear distinction between rational and irrational concepts within the realm of herbalism. Charles chose the irrational option of not differentiating between the two which, unsurprisingly, was less than successful:

- His plan for statutory regulation of traditional herbalists has failed.
- UK research activity in herbal medicine is negligibly small (~1-2 % of the global output).
- Funds for research into herbal medicine in the UK have remained unavailable.
- The NHS has ignored herbalism almost completely.
- Traditional herbalism remains unsupported by evidence.
- Traditional herbalism is practised by merely ~700 herbalists in the UK.[123]
- Rational phytotherapy is hardly practised at all in the UK.

In other words, despite Charles' relentless and long-standing efforts, UK herbal medicine is far from being successful.

Box 15

Examples of herbal medicines that are supported by at least some reasonably sound evidence

- Devil's claw (pain)
- Echinacea (common cold)
- Garlic (hypertension, hyperlipidaemia)
- Horse chestnut (varicose veins)
- Red clover (menopausal symptoms)
- St John's wort (depression)
- Willow (pain)

Sixteen

The Smallwood Report

In October 2005, a report was launched, i.e. handed to politicians in the House of Parliament, entitled 'The Role of Complementary and Alternative Medicine in the NHS'.[124] Its lead author was the economist Christopher Smallwood, with no previous experience in healthcare-related subjects; he had been commissioned directly by Charles to write the document which became better known as the 'Smallwood Report'.

16.1. Charles' intention

The stated purpose of the report was: '…to examine evidence relating first to the effectiveness and then to the associated costs of mainstream complementary therapies'.[124] The report was never formally peer-reviewed and the intent was never to publish it in order to generate a well-informed discussion; it was purely for influencing UK health politicians.

In 2000, Charles had made a passionate plea for more funding of clinical trials of alternative medicine (Chapter 11). By 2005, many more trials of alternative medicine (I estimate about 3,000) from across the world had emerged. Their results were not what Charles had hoped for. The scientific evidence supporting alternative medicine was generally speaking unimpressive (in fact, it is so to the present day) and, crucially, it was getting less positive as time passed. Specifically, there was no good evidence to suggest that any alternative therapy would meet the efficacy criteria for getting adopted by the NHS. The prospect of using more alternative medicine was therefore further remote than ever. Charles' dream of integrated medicine (Chapter 13) appeared less and less realistic.

In view of this desperate situation, a new idea had been generated: if one could show that the integration of some alternative therapies into the NHS would save significant sums of money, politicians might be persuaded to follow his dream. Never mind the evidence on efficacy, the financial argument might still win the day.

Smallwood thus calculated the savings that an integration of selected alternative therapies would achieve for the NHS. He received help from a commercial group called 'Freshminds'[125] and was asked to write a report, and present it to politicians at its launch in the House of Parliament. A spokeswoman of Charles later commented: 'By commissioning this report the Prince hoped to further encourage an informed debate about how an evidence-based integrated approach to health, which draws on the best of both orthodox and complementary medicines, might offer wider benefits.'[126]

This, of course, begs the question, why was the report not submitted to any kind of scrutiny such as peer-review?

16.2. The evidence

Mr Smallwood visited me personally in Exeter and tried to involve my expertise. When I saw the many serious errors in his draft document, I offered to help and correct them. This would have been relatively easy, because I had just completed a similar investigation for the World Health Organisation (Box 16).[127] When my offer was refused, I asked for my name to be withdrawn from the document. Shortly before the report was due to be launched, a journalist from *The Times* asked me for a comment, and I criticised Smallwood's methodology without disclosing the report's findings (which was not necessary, as the report had been leaked to *The Times*). The following day, this was the title story of *The Times*.[128] Wikipedia correctly sums up what followed: 'As a dire consequence of Ernst's response, Charles' secretary Sir Michael Peat filed a complaint to Exeter University to initiate an investigation file on Ernst for misconduct. Although Ernst was exonerated, his department was disbanded due to stoppage of funding and he was forced into early retirement. The role of Prince Charles' royal office has been scrutinized as a result of the interference.'[129] A full account of these events can be found in my memoir.[5]

Days after the publication of the Smallwood Report, Prince Charles was quoted as saying that he did not promote alternative medicine 'because of some self-indulgent pet projects, or because of some half-baked obsession with unsubstantiated quackery'.[130] Yet, this was precisely how many rational thinkers interpreted the Smallwood Report. Richard Horton, editor of the *Lancet*, for instance, did not mince his words:

> …Let's be clear: this report contains dangerous nonsense.
>
> The summary includes the following: 'The best evidence for homeopathy, in terms both of improved health benefits and reduced costs, is associated with its use as an alternative to conventional medicine in relation to a number of everyday conditions, particularly asthma.'
>
> About 1,400 people die from asthma every year in the UK. It is a life-threatening condition that can be controlled by the effective use of drugs. The idea that homeopathy can replace conventional treatment, as the prince's report suggests, is absolutely wrong. Not one shred of reliable evidence exists to support this incredibly misjudged claim. Lives will be lost if this practice, apparently endorsed by the president of the GMC, is followed.
>
> Is this what listening to patients has come to mean? We are losing our grip on a rational scientific medicine that has brought benefits to millions, and which is now being eroded by the complicity of doctors who should know better and a prince who seems to know nothing at all.
>
> Dr Richard Horton
> Editor, The Lancet[131]

Any critical analysis of the report would have spotted numerous flaws and inaccuracies. A few examples will have to suffice:

- '...manipulative therapies offer advantages over conventional treatments for lower back pain, particularly acute pain.' A Cochrane review (such papers are widely considered the most reliable evidence available), however, concluded at the time that 'there is no evidence that spinal manipulative therapy is superior to other standard treatments...'[132]
- Viral infections such as the common cold were, according to Smallwood, treatable with Echinacea extract. The most conclusive series of trials, however, showed little benefit in terms of treatment or prevention.[133]
- Homoeopathy was recommended as a cost-saving treatment for asthma. A Cochrane review, however, fails to demonstrate efficacy.[134]
- The widespread use of homoeopathy, it was claimed, would save $4 billion on the national prescription drugs bill. Yet, this notion was based not on controlled data but on a 'case study'. About a dozen systematic reviews/meta-analyses failed to confirm that homeopathic remedies are more than a placebo.[135]
- '...economy-wide benefits running into the hundreds of millions of pounds could result if a significant reduction in the time off work associated with lower back pain alone could be achieved as a result of the wider application of complementary/alternative (CAM) therapies.' No conclusive data were provided to suggest that CAM could reduce absenteeism due to back problems.
- '...a number of CAM treatments offer the possibility of significant savings in direct health costs.' The best evidence available at the time did not support this statement.[125,136]
- '...if 4% of GPs were to... [offer homeopathy] a large saving (£190 million) would result.' No conclusive data were provided to support this statement.

The Smallwood Report became the subject of numerous newspaper articles in the UK and abroad. Some, including myself, argued that Charles might have been over-stepping the boundaries of his constitutional role by trying to influence UK health politics. Mr Smallwood had told me that the money for his report had been donated by Dame Shirley Porter, the heir to the Tesco empire, who had been described as 'the most corrupt British public figure in living memory'.[137] She was once ordered to repay £42m for her 'blatant and dishonest misuse of public power'.[138]

Headed by Professor Baum, a group of leading UK medical experts eventually published an open letter in *The Times* to the chief executives of the NHS and Primary Care Trusts:

> Re Use of 'alternative' medicine in the NHS
> We are a group of physicians and scientists who are concerned about ways in which unproven or disproved treatments are being encouraged for general use in the NHS. We would ask you to review practices in your own trust, and to join us in representing our concerns to the Department of Health because we want patients to benefit from the best treatments available.
>
> There are two particular developments to which we would like to draw your attention. First, there is now overt promotion of homeopathy in parts of the NHS (including the NHS Direct website). It is an implausible treatment for which over a dozen systematic reviews have failed to produce convincing evidence of effectiveness. Despite this, a recently-published patient guide, promoting use of homeopathy without making the lack of proven efficacy clear to patients, is being made available through government funding. Further suggestions about benefits of homeopathy in the treatment of asthma have been made in the Smallwood Report and in another publication by the Department of Health designed to give primary care groups 'a basic source of reference on complementary and alternative therapies.' A Cochrane review of all relevant studies, however, failed to confirm any benefits for asthma treatment.
>
> Secondly, as you may know, there has been a concerted campaign to promote complementary and alternative medicine as a component of healthcare provision. Treatments

covered by this definition include some which have not been tested as pharmaceutical products, but which are known to cause adverse effects, and others that have no demonstrable benefits. While medical practice must remain open to new discoveries for which there is convincing evidence, including any branded as 'alternative', it would be highly irresponsible to embrace any medicine as though it were a matter of principle.

At a time when the NHS is under intense pressure, patients, the public and the NHS are best served by using the available funds for treatments that are based on solid evidence. Furthermore, as someone in a position of accountability for resource distribution, you will be familiar with just how publicly emotive the decisions concerning which therapies to provide under the NHS can be; our ability to explain and justify to patients the selection of treatments, and to account for expenditure on them more widely, is compromised if we abandon our reference to evidence. We are sensitive to the needs of patients for complementary care to enhance well-being and for spiritual support to deal with the fear of death at a time of critical illness, all of which can be supported through services already available within the NHS without resorting to false claims.

These are not trivial matters. We urge you to take an early opportunity to review practice in your own trust with a view to ensuring that patients do not receive misleading information about the effectiveness of alternative medicines. We would also ask you to write to the Department of Health requesting evidence-based information for trusts and for patients with respect to alternative medicine.

Yours sincerely

Professor Frances Ashcroft FRS
University Laboratory of Physiology, Oxford
Professor Sir Colin Berry

Emeritus Professor of Pathology, Queen Mary, London
Professor Gustav Born FRS
Emeritus Professor of Pharmacology, Kings College London
Professor Sir James Black FRS
Kings College London
Professor David Colquhoun FRS
University College London
Professor Peter Dawson
Clinical Director of Imaging, University College London
Professor Edzard Ernst
Peninsula Medical School, Exeter
Professor John Garrow
Emeritus Professor of Human Nutrition, London
Professor Sir Keith Peters FRS
President, The Academy of Medical Sciences
Mr Leslie Rose
Consultant Clinical Scientist
Professor Raymond Tallis
Emeritus Professor of Geriatric Medicine, University of Manchester
Professor Lewis Wolpert CBE FRS
University College London.[139]

16.3. Consequences

In the end, the Smallwood Report was politely acknowledged by the politicians for whom it had been written. Whether they realised how flawed it was is unknown. What is evident, however, is that it remained without any noticeable influence on the use of alternative medicine in the NHS.

Nonetheless, one consequence of the Smallwood Report is undeniable: an official complaint by Charles' first private secretary to my vice-chancellor, Professor Steve Smith, alleging that I had violated the rules of confidentiality, led to the closure of my research unit.[5] Instead of supporting research into alternative medicine, as he so often claimed, Charles (directly or indirectly) had managed to close the leading research centre in alternative medicine[140] that had published well over 1,000 papers (many of them not confirming Charles' beliefs) in the peer-reviewed medical literature.

Box 16

Our economic evaluation of complementary and alternative medicine for the WHO[127]

- We located all relevant studies available at the time.
- A total of 28 such investigations were found and summarised.
- They related to a diverse range of alternative therapies.
- They used different methods of economic evaluation.
- The totality of the evidence did not show that integration of alternative medicine would save money for healthcare systems.

Seventeen
World Health Organisation

In 2006, Charles took his alternative medicine agenda on to the international stage by delivering a lecture to the general assembly of the World Health Organisation (WHO) in Geneva. Here are some key passages from his speech, to which I have added [in square brackets] references that relate to my comments below.[141]

17.1. Charles' lecture

...For the past 24 years I have argued that patients should be able to gain the benefit of the 'best of both worlds'—complementary and orthodox—as part of an integrated approach to healing. Many of today's complementary therapies are rooted in ancient traditions that intuitively understood the need to maintain balance and harmony with our minds, bodies and the natural world. Much of this knowledge, often based on oral traditions, is sadly being lost[1] yet, orthodox medicine has so much to learn from it. It is tragic, it seems to me—and indeed to many people who have studied this whole area—that in the ceaseless rush to 'modernize', many beneficial approaches, which have been tried and tested[2] and have shown themselves to be effective,[3] have been cast aside because they are deemed to be 'old-fashioned' or 'irrelevant' to today's needs.[4]

There are clear examples which come to mind, particularly in the fields of acupuncture[5] and herbal medicines.[6] While scientists try to learn more about how acupuncture works, increasingly robust evidence, drawn from a number of international studies, indicates that it does work, particularly for the treatment of conditions such as osteoarthritis of the knee.[7] It can, according to the evidence, also alleviate the nausea and vomiting that can be so debilitating for those taking anti-cancer drugs.[8]

In the case of herbal applications such as St John's Wort (Hypericum perforatum), which has been used since the time of the ancient Greeks, about thirty clinical trials have shown some positive effects in treating non-severe depression, with a remarkably low incidence of side-effects…[9]

Hippocrates said 'First, do no harm'.[10] I believe that the proper mix of proven complementary, traditional and modern remedies, which emphasizes the active participation of the patient can help to create a powerful healing force for our world…[11]

In the United Kingdom, my Foundation for Integrated Health has been the leading champion of this integrated approach for the past eleven years.[12] Another of my organizations, the International Business Leaders Forum, has been working with the W.H.O. on a number of projects aimed at, amongst other things, finding ways of improving health through better diets and increasing physical activity, in a number of countries…

…The case of Artemesia is a classic example of where real progress can be made. A naturally growing plant, long used in China for treating Malaria,[13] Artemesia is now a treatment of choice in many parts of the World…

…Could I suggest that each country represented here today looks at the possibility, over the next five years say, of developing its own integrated plan for future health and care, perhaps beginning with a pilot or feasibility study…? If I may say so, such a plan would reflect your disparate cultures and medical traditions and would recognize the importance of all aspects of the natural environment. It would be a plan that would integrate medical services with individual and community approaches to health and self-care; a plan that might build upon current examples of integrated health and care, which exist everywhere.[14] And if you ever get round to devising such a plan, why not ask your Finance ministers to quantify the savings from this new and emphatic focus on prevention as well as cure?[15]

You might be interested to know on that score that last year I commissioned a report in order to encourage a better informed debate about the effectiveness of different therapies and treatments which might eventually result in savings. The report, compiled by a British economist, Christopher Smallwood, was published last October and it found evidence that complementary approaches could help to fill gaps in some orthodox treatments, particularly in relation to many chronic conditions such as lower back pain, osteoarthritis of the knee, stress, anxiety and depression, and post-operative nausea and pain.[16] I am not here to tell you what to do, but I would merely suggest that you might find a similar approach helpful.[17]

17.2. My comments

Here are my brief comments related to the references 1–17 inserted by me into Charles' speech:

1. Traditional knowledge in healthcare is rarely lost (and the appeal to tradition is a logical fallacy[142]). Scientists regularly investigate traditional knowledge and, if it seems remotely plausible, they test it in clinical trials. If these tests turn out positive, they would consider it as a possible treatment. An example is bloodletting, which was used for centuries as a panacea. When it was finally submitted to scientific tests, it was shown to be harmful and swiftly abandoned. A more encouraging example is St John's Wort, which had long been used for various diseases. When it was scientifically tested, it turned out to be effective for depression, a condition for which it was not traditionally employed.[143] Today, it is widely used for this condition.

2. This seems to be a contradiction to what Charles stated just now, namely that this knowledge gets lost.
3. Most of the approaches Charles is speaking of have not been shown to be effective.
4. Ancient methods are never cast aside because they are old-fashioned or irrelevant. They are cast aside if they lack plausibility and/or efficacy and/or safety.
5. For the year 2006, Medline, the largest databank for medical publications, lists more than 1,000 articles on acupuncture. Despite this abundance of research, the effectiveness of acupuncture remains an unresolved issue.[144]
6. For 2006, Medline lists ~1,300 articles on herbal medicine, but only very few herbal remedies had been shown to be a truly effective treatment for any condition.
7. This is true: in 2006, the evidence for acupuncture as a treatment of knee osteoarthritis was positive.[145] Yet, the issue is more complex: osteoarthritis of the knee usually needs surgery; acupuncture will at best reduce the pain and delay surgery (the latter effect is not necessarily desirable).
8. The evidence that acupuncture can alleviate nausea and vomiting caused by chemotherapy was never strong.[146] More crucially, modern drug treatments of this condition might be more acceptable, practical, and effective than acupuncture. Charles seems to be focused on alternative options, while forgetting that better methods have long been identified through what he calls the ceaseless rush to modernise. This, of course, begs the question whether Charles truly wants the best for patients or whether he is more interested in pushing his alternative medicine agenda.
9. St John's Wort is indeed one of the best-researched alternative remedies. However, one should mention that, through powerful interactions with many prescribed medications, it can cause serious, even life-threatening, adverse effects. These problems were known since the late 1990s.[147]
10. Many people think so; yet the famous oath does not contain this sentence (Box 17).[148]

11. What is the 'proper' mix? Most people would say that it must focus on treatments that are soundly backed by evidence. From what Charles states, we have to assume that he does not agree with this view.
12. See Chapter 10.
13. It is true that Artemesia was used in Traditional Chinese Medicine (TCM). It is, however, not true that it was used for malaria.[149] The discovery that it is effective for malaria is yet another consequence of the ceaseless rush to modernise which Charles seems to despise.[150]
14. Charles seems to think that all the international research he wants to initiate here is bound to yield positive results. He does not seem to consider that research is not for confirming his belief but for testing hypotheses which, of course, can (and often do) turn out to be wrong.
15. Charles seems unable to imagine that alternative medicine might be an extra expense that would increase the cost of healthcare (Chapter 16).
16. See Chapter 16.
17. In the UK, the Smallwood Report had just caused a major embarrassment for Charles, yet here he recommends the same to all other nations of the world.

These comments might seem like an exercise in nit-picking. Arguably, the UK's most authoritative source made a less detailed but possibly more devastating criticism. Professor David Read, vice-president of the Royal Society, said: 'We share the concerns that some treatments labelled as complementary and alternative medicines have not been properly tested and are known to cause adverse effects, while others have no demonstrable benefits. We also support the view that patients should not receive misleading information about the effectiveness of complementary medicine.'[151]

17.3. Consequences

Even though Charles' speech at the WHO was widely reported, it had no significant effects in the UK or internationally. Seven years after Charles' speech, the WHO reported: 'More countries have gradually come to accept the contribution that traditional and complementary medicine can make to the health and well-being of individuals and to the comprehensiveness of their health care systems... In general, data reported by Member States show that progress in matters related to the regulation of traditional and complementary medicine products, practices and practitioners is not occurring at an equal pace.'[152]

Box 17

The Hippocratic Oath

- Contrary to a common belief, most medical schools do not ever require graduates to take the oath.
- The actual text of the oath is no longer applicable to modern times.
- The oath does not contain the famous sentence 'first do no harm'.[153]
- Doctors do harm all time, for instance, when they carry out a painful procedure.
- The ethical imperative is therefore not 'first do no harm', but 'always do more good than harm' and obtain fully informed consent (an obligation that is widely neglected in alternative medicine).[153]

Eighteen
Traditional Chinese Medicine

Traditional Chinese Medicine (TCM) is the umbrella term for a range of modalities that had been used in ancient China. It is a construct invented by Mao Zedong, who lumped all historical Chinese treatments together under the TCM umbrella and created the 'barefoot doctor' to practice TCM nationwide—not because he believed in TCM (he actually didn't), but because China was desperately short of doctors and needed at least a semblance of healthcare.

18.1. Charles' view

Charles has only rarely commented on TCM. He did, however, regularly speak about aspects of TCM, such as acupuncture (Chapter 17 and 25). In 2007, the People's Republic of China recorded the visit of Fu Ying, its ambassador in London at the time, to Clarence House and reported that Charles used that occasion to praised TCM. 'He hoped that it could be included in the modern medical system… and was willing to make a contribution to it.'[154]

18.2. The evidence

In recent years, TCM has become hugely popular outside China. It earns about US$50 billion per year for China's economy. Therefore, TCM is a financially and ideologically important issue for the country.[155] Even President Xi Jinping has thrown his weight behind TCM, which he says is crucial to the development of China's healthcare.[156] Criticism of TCM is not permitted in China and is harshly suppressed.[157]

Even though TCM modalities differ in many respects, they are claimed to have in common that they are based on assumptions which originate from Taoist philosophy:

- The human body is a miniature version of the universe.
- Harmony between the two opposing forces, yin and yang, means health.
- Disease is caused by an imbalance between these forces.
- Five elements—fire, earth, wood, metal, and water—symbolize all phenomena and explain the functions of the body.
- The flow of vital energy, qi or chi, through the body in meridians is essential for maintaining health.

These assumptions are not in keeping with the known facts about physiology, anatomy, etc. One prominent critic of TCM rightly cautioned: 'TCM is a pre-scientific superstitious view of biology and illness, similar to the humoral theory of Galen, or the notions of any pre-scientific culture. It is strange and unscientific to treat TCM as anything else. Any individual diagnostic or treatment method within TCM should be evaluated according to standard principles of science and science-based medicine, and not given special treatment.'[158]

Considering Charles' often-voiced concerns for the environment and nature, it is relevant to mention that many traditional remedies used in TCM contain animal parts. Not only is there no evidence that such ingredients have any positive health effects, they also threaten the survival of several endangered species.

In general, the evidence on TCM is neither consistently positive nor trustworthy (Box 18). In particular, there is no evidence to show that the individualised approach of Chinese herbalists (Chapter 15) is effective.[119] China's drug regulator records more than 230,000 reports of adverse effects from TCM each year, and Chinese herbal medicines carry multiple direct risks:

- One or more ingredients can be toxic.[159]

- Some herbal remedies have been shown to be contaminated with toxic materials such as heavy metals.[121]
- Others are adulterated with synthetic drugs such as steroids.[119]
- Others again can interact with prescription drugs taken concomitantly.[160]
- The risk for endangered species.[161]

To this, we have to add the indirect risk of employing useless treatments for otherwise treatable conditions. In view of all this, Charles' (indirect) endorsement of TCM practices seems more than a little surprising.

18.3. Consequences

Despite some support from Charles, concern over TCM is growing. Criticism is being voiced more and more openly and urgently.[162] The Federation of European Academies of Medicine (FEAM) and the European Academies' Science Advisory Council, for instance, issued a statement in 2019 that stressed many valuable points about the glaring deficits of TCM, including the following:[163]

- 'There should be consistent proof underlying the regulatory requirements for scrutiny to demonstrate efficacy, safety and quality for all products and practices for human medicine. There must be verifiable and objective evidence, commensurate with the nature of the claims being made. In the absence of such evidence, a product should be neither approvable nor registrable by national regulatory agencies for the designation medicinal product...'
- 'Diagnostic procedures should also be evidence-based and include validated diagnostic instruments to provide objective, reliable, reproducible assessment and reduce inter-rater variability. Whatever the diagnostic approach utilised, practitioners should be appropriately trained and audited by professional bodies.'

- 'Similarly, use of other TCM procedures such as acupuncture should be evidence-based to demonstrate efficacy and safety, and subject to professional standards.'
- 'Evidence-based public health systems and medical insurance systems should not reimburse products and practices unless they are demonstrated to be efficacious and safe by rigorous pre-marketing testing: a robust evidence base is essential for all medicines.'
- 'The composition of standardised TCM remedies should be labelled in a similar way to other health products. That is, there should be an accurate, clear, verifiable and simple description of the ingredients and their amounts present in the formulation. TCM diagnostic and therapeutic procedures should, likewise, be clearly explained in patient information literature.'
- 'Advertising and marketing of TCM products and services must conform to established standards of accuracy and clarity. Promotional claims for efficacy, safety and quality should not be made without demonstrable and reproducible evidence.'

Despite Charles' support for some TCM practices, its availability on the NHS remains very limited and does not seem to be increasing.

Box 18

The trustworthiness of TCM research originating from China

- Research into TCM is highly active in China and often seems more promotional than scientific.
- Its methodological quality is often poor.
- Almost all Chinese TCM studies (at least up to the start of the new millennium) reported positive results.[164]
- There is evidence to suggest that many Chinese investigations are not reliable.[165]
- One survey of Chinese clinical trials concluded that the results of more than 80 percent of these studies are 'fabricated'.[166]

Nineteen

The 'GetWellUK' Study

From February 2007 to February 2008, an organisation by the name of 'GetWellUK' (Box 19) conducted the UK's first government-backed pilot study of alternative medicine. Sixteen practitioners provided treatments including acupuncture, osteopathy, and aromatherapy to more than 700 patients at two GP practices in Belfast and Derry.

19.1. Charles' intentions

The study was designed to demonstrate the benefit of integrating alternative medicine into the routine of the NHS. In 2014, *BBC News* published the following information about the background to the study:

> Prince Charles has been a well-known supporter of complementary medicine. According to another former Labour cabinet minister, Peter Hain, it was a topic they shared an interest in. 'He had been constantly frustrated at his inability to persuade any health ministers anywhere that that was a good idea, and so he, as he once described it to me, found me unique from this point of view, in being somebody that actually agreed with him on this, and might want to deliver it.' Mr Hain added: 'When I was Secretary of State for Northern Ireland in 2005–7, he was delighted when I told him that since I was running the place I could more or less do what I wanted to do. I was able to introduce a trial for complementary medicine on the NHS, and it had spectacularly good results, that people's well-being and health was vastly improved. And

when he learnt about this he was really enthusiastic and tried to persuade the Welsh government to do the same thing and the government in Whitehall to do the same thing for England, but not successfully,' added Mr Hain.[167]

In his book *Harmony*, Charles provided his interpretation of the study's findings:

> Following persistent encouragement from my Foundation for Integrated Health, a modest one-year pilot scheme took place... Overall the results showed how both patients and the health service can benefit from more integrated approaches to healthcare... This, to me, is another example of how a joined-up approach, based on the principles of harmony, can help us to do better for people and cut excess costs...[2]

19.2. The evidence

The UK Department of Health published some evidence about this study:

> Aims and Objectives
> The aim of the project was to pilot services integrating complementary medicine into existing primary care services in Northern Ireland. GetWellUK provided this pilot project for the Department for Health, Social Services and Public Safety (DHSSPS) during 2007.
> The objectives were:
> • To measure the health outcomes of the service and monitor health improvements.
> • To redress inequalities in access to complementary medicine by providing therapies through the NHS, allowing access regardless of income.
> • To contribute to best practise in the field of delivering complementary therapies through primary care.
> • To provide work for suitably skilled and qualified practitioners.
> • To increase patient satisfaction with quick access to expert care.

- To help patients learn skills to improve and retain their health.
- To free up GP time to work with other patients.
- To deliver the programme for 700 patients.

Results

The results of the pilot were analysed by Social and Market Research, who produced this report.[168]

Following the pilot, 80% of patients reported an improvement in their symptoms, 64% took less time off work and 55% reduced their use of painkillers.

In the pilot, 713 patients with a range of ages and demographic backgrounds and either physical or mental health conditions were referred to various complementary and alternative medicine (CAM) therapies via nine GP practices in Belfast and Londonderry. Patients assessed their own health and wellbeing pre and post therapy and GPs and CAM practitioners also rated patients' responses to treatment and the overall effectiveness of the scheme.

Health improvement
- 81% of patients reported an improvement in their physical health
- 79% reported an improvement in their mental health
- 84% of patients linked an improvement in their health and wellbeing directly to their CAM treatment
- In 65% of patient cases, GPs documented a health improvement, correlating closely to patient-reported improvements
- 94% of patients said they would recommend CAM to another patient with their condition
- 87% of patient indicated a desire to continue with their CAM treatment

Painkillers and medication
- Half of GPs reported prescribing less medication and all reported that patients had indicated to them that they needed less
- 62% of patients reported suffering from less pain

- 55% reported using less painkillers following treatment
- Patients using medication reduced from 75% before treatment to 61% after treatment
- 44% of those taking medication before treatment had reduced their use afterwards

Health service and social benefits
- 24% of patients who used health services prior to treatment (i.e. primary and secondary care, accident and emergency) reported using the services less after treatment
- 65% of GPs reported seeing the patient less following the CAM referral
- Half of GPs said the scheme had reduced their workload and 17% reported a financial saving for their practice…
- Half of GPs said their patients were using secondary care services less.

Given the evidence of health gain documented by patients, GPs and CAM practitioners, it is recommended that DHSSPS and the project partners explore the potential for making CAM more widely available to patients across Northern Ireland. Not only has this project documented significant health gains for patients, but it has also highlighted the potential economic savings likely to accrue from a reduction in patient use of primary and other health care services, a reduction in prescribing levels and reduced absenteeism from work due to ill health.[168]

To lay people, this might look impressive. Yet, from a scientific point of view, the study is an embarrassment. Its multiple aims cannot possibly be met by a 'pilot study' (the aim of a pilot study is to test a protocol for a later definitive trial). The study has too many flaws to mention all here. A small selection should suffice:

- The results do not show that the 'health gains' were caused by the alternative therapies.
- They also do not reveal a potential for economic savings caused by the interventions.

- As there was no control group, it is not possible to attribute any of the outcomes to the therapies offered. They could have been due to a range of factors that are unrelated to the alternative therapies (e.g. placebo effects, natural history of the disease, regression towards the mean).
- Most outcome measures were not objectively verified.
- The patients were self-selected which means that their expectation of a benefit accounts for much of the observed result.
- All patients had conventional treatments in parallel. This further complicated the attribution of a cause to the effects.

The methodology of this study was of such exceptionally poor quality that its findings never got published in a peer-reviewed journal. The scientific community only had pitiful smiles or sarcastic scorn for this 'trial':

> I see this as nothing short of an attempted fraud to extract NHS money for traders in quackery... What is quite remarkable about this so called study is that the money to conduct the trial was given to a lobby group for promoting the inclusion of alternative medicine in the NHS. It is difficult to imagine any other area of government where a group with large vested interests was given permission to promote their business, under the guise of science, using tax payers money. Independent, this report is not. Peter Hain, the then Northern Ireland Secretary and supporter of quackery, gave the money (£200,000) to an outfit called GetWellUK. GetWellUK, run by Boo Armstrong, is a private company specifically set up to be, in their words, 'the best supplier of complementary healthcare to the National Health Service.' Only a fool would think any dispassionate appraisal would come out of such a lobby group.[169]

And Professor David Colquhoun FRS, pointed out that the project was a farce:

At the end of the 'pilot scheme' there will have been no proper assessment of the effectiveness of the treatments. The analysis has come from a market research company called SMR... It is not a scientific analysis. It is at best a customer satisfaction survey. At worst, it is a set of graphs and figures that will please SMR's clients — GetWellUK — so that they can use it for misleading PR.[170]

19.3. Consequences

The flaws of the GetWellUK pilot project were obvious and fundamental for anyone with a minimum of understanding of science. Yet, Charles felt the pilot was a resounding success and lobbied health politicians to repeat the exercise in England.[116] Thankfully, this never happened.

Boo Armstrong, the founder of GetWellUK who had been in charge of the pilot study, was subsequently appointed as 'Medical Director' of Charles' Foundation for Integrated Health. In the end, neither Northern Ireland nor any other part of the UK implemented the programme, and the whole pseudoscientific exercise was all but forgotten.

Box 19

GetWellUK

- GetWellUK was a private, not for profit organisation.
- It was founded by the late Boo Armstrong (1974–2012) in 2004.
- GetWellUK was allegedly established in response to the House of Lords report of 2000 (Box 11).[170]
- Armstrong's mission was to make GetWellUK 'the best supplier of complementary healthcare to the NHS'.[171]
- The pilot study in Northern Ireland was financed by the UK government at the expense of the UK taxpayer.
- After Boo died aged 37, GetWellUK closed.

Twenty

Bravewell

The Bravewell Collaborative was an initiative of leading US philanthropists who collaborated with the aim to transform health care in the US and improve the health of the American public through the advancement of integrated medicine (Chapter 13). The specific goal of the group was to fund strategic programmes and to help the field of integrated medicine further its work in providing some of the solutions to the US healthcare system's problems.[172]

20.1. Charles' involvement

In 2008, Charles signed an agreement to 'establish a partnership with the Bravewell Collaborative focused on improving the health of the public in both countries by advancing the use of integrated health'.[173] A few months later, John and Christy Mack, Bravewell's founders, hosted a ceremony to honour some of those alternative practitioners whom they had supported. Charles contributed via video link to this event stating that he was looking forward to working across the Atlantic between his Foundation for Integrated Health and Bravewell in the US.[174]

20.2. The evidence

'The practice of medicine should be focused with an emphasis on the interconnectedness of mind, body, spirit and community', said Christy Mack, a Reiki practitioner and co-founder of the Bravewell Collaborative.[175] 'We want Bravewell to be a catalyst for change, and we think we are reaching the tipping point', said

Mack, stressing that she won't be satisfied until the medical establishment accepts her point of view. Her husband John Mack, a Wall Street titan, said he has been 'moved by the passion and dedication of Christy's doctors to improving people's health by treating the whole person, not just the disease' by means of drugs and surgery.[175]

The Bravewell Collaborative was aimed at university medical centres and financed the construction of integrative health facilities. It funded alternative medicine research, supported the development of the curricula on the topic for medical schools, and trained 30 doctors a year in integrative medicine at a cost of $30,000 per doctor, using the facilities of Dr Andrew Weil (Box 20). The organisation also gave a biannual $100,000 prize to recipients of its 'Bravewell Leadership Award', given to doctors who have made a big impact within the field.

Bravewell had considerable success. In 2001, there were just six integrated medicine facilities at US medical schools. Less than a decade later, there were 36. In 2004, the Bravewell Collaborative formed the 'Bravewell Clinical Network', a group of eight integrative medicine centres across the US. The network trained clinical fellows and spread the word about integrated care.

Based in Minneapolis with a tiny staff of just six people, the Bravewell Collaborative raised $21 million in contributions and pledges. 'Our dream is that Bravewell's vision will be so successful that the need for Bravewell to exist will no longer be necessary. That's an unusual goal — to be so good that you don't have to exist anymore', stated Mrs Mack.[174]

Even if one disagreed with its aims, one has to concede that the Bravewell Collaborative did achieve its aims. 'Arguably no single private organization has been more effective at promoting the infiltration of CAM/IM into medical academia (or, as I like to call it, quackademic medicine) than the Bravewell Collaborative', wrote the prominent US oncologist and critic of integrated medicine, David Gorski.[176] The success of the Bravewell Collaborative was such that they considered their main goals to have been reached and, in 2015, Christy Marks announced the closure of initiative:

Dear Friends and Colleagues:

This letter is both a farewell and a thank you. From the beginning, our goal was to not exist. So on June 17, 2015 The Bravewell Collaborative will close its doors with pride in our accomplishments and immense gratitude for all the people and institutions with whom we worked along the way.

We formed back in 2002, and have since worked together to change how Americans think about their health and the kind of healthcare they receive, and to bring about the cultural change necessary to create a healthier nation. Then, when our principal strategies had achieved our goals, and when integrative medicine had become part of the national conversation on healthcare, our members collectively decided that it was time to sunset the organization.

But needless to say, none of this great work could have been possible without you—our friends, our colleagues, our partners, and our supporters. Together, we made huge strides in making prevention, patient empowerment, and healthy living part of the nation's priorities. And together, we developed the core template for care that addresses not just the body, but also the patient's mind and spirit. The ranks have swelled— thousands of others are carrying the torch forward and certainly many of us will continue to fund initiatives from our private foundations.

From all of us at Bravewell, thank you for being a part of this effort.

With my best regards,
Christy Mack[177]

To cite Dr Gorski again:

The problem, once again, is that it is not necessary to 'integrate' pseudoscience with science-based medicine (SBM) in order to practice collaborative patient-centered care. It's really not, just as it's not necessary to 'integrate' pseudoscience into SBM in order to be 'holistic' or to 'take care of the whole patient.' Bravewell keeps selling that false dichotomy. I'm not buying, and neither is any member of the SBM team... As our very own Harriet Hall has so frequently and eloquently said about naturopaths and CAM/IM practitioners, 'What they do that is good is not special, and what they do that is special is not good.'[176]

20.3. Consequences

Charles' attempt to learn and benefit from the considerable drive of his like-minded American friends resulted in nothing. A meaningful cooperation between Charles and Bravewell never emerged.

Yet, for judging Charles' various activities in promoting alternative medicine, the history of the Bravewell Collaborative is relevant. However one feels about integrative medicine, it shows that with a clear strategy and direct funding of research and researchers even the most ambitious aims can be reached. Charles' Foundation for Integrative Health (Chapter 10) never did fund research directly and looked unprofessional in comparison. Consequently, its impact remained insignificant.

Box 20

Andrew Weil

- Andrew Weil (born 1942) has a medical degree from Harvard.
- He is one of America's most prominent advocates of alternative medicine.
- In 1994, Weil founded what today is the 'Andrew Weil Center for Integrative Medicine' at the University of Arizona, Tucson, USA.
- He is well-known for his experiments with several mind-altering substances.
- Weil owns several well-earning commercial enterprises and authored many best-selling books.

Twenty-One

Duchy Originals Detox Tincture

Duchy Originals is a limited company selling organic food. It was set up by Charles in 1990 and named after the Duchy of Cornwall estates that are held in trust by him and from which he derived his annual income.[178] Waitrose Duchy Organics, as the company is called today, operates entirely separately from the Duchy of Cornwall.[179] In 2008, Charles surprised the world by launching three herbal remedies, including the 'Duchy Herbals Detox Tincture', produced in partnership with the homeopathic pharmacy 'Nelsons'[180] and marketed via the Duchy Originals brand.

21.1. Charles' intention

The Duchy Herbals Detox Tincture was marketed as a food supplement. It combined extracts of artichoke and dandelion and promised to 'rid the body of toxins while aiding digestion'. Andrew Baker, chief executive of Duchy Originals, claimed that 'it is a natural aid to digestion and supports the body's natural elimination processes. It is not—and has never been described as —a medicine, remedy or cure for any disease. Duchy Herbals Detox Tincture contains globe artichoke and dandelion, which both have a long history of traditional use for aiding digestion.'[181]

21.2. The evidence

Detox is short for 'detoxification'. In conventional medicine, the term is used for treatments that wean drug-dependent patients off their drugs. In alternative medicine, detox has become an umbrella term for a wide range of treatments that allegedly rid our bodies of toxins.[29] Charles' Detox Tincture evidently belonged to the latter category.

The belief that our body is full of toxins which threaten our good health is extremely widespread across all areas of alternative medicine. These toxins can allegedly originate from our body's own metabolism, from the environment, from prescribed drugs, or from the food we consume. Proponents of detox claim that their treatments help the body get rid of these toxins and thus restore health. The nature of the toxins in question is rarely defined by proponents of detox. Thus, we cannot easily determine whether the treatments are successful or not in eliminating them from the body.[29]

The notion of 'detox' is based on the naturopathic notion that: 'toxins damage the body in an insidious and cumulative way. Once the detoxification system becomes overloaded, toxic metabolites accumulate, and sensitivity to other chemicals, some of which are not normally toxic, becomes progressively greater. This accumulation of toxins can wreak havoc on normal metabolic processes.'[182] The assumption that some toxins can accumulate in our body is, of course, correct. However, it is usually exaggerated beyond all proportions by detox enthusiasts. Crucially, the assumption that any of the alternative detox treatments effectively eliminate toxins from our body lacks sound evidence. Our body has powerful mechanisms to achieve this aim on its own; when they fail, we fall seriously ill and the last thing we need then is detox. Something all these treatments have in common is that there is no evidence that they eliminate anything other than some cash from consumers.[183] Certainly, the Duchy Originals Detox Tincture had not been tested for its ability to eliminate toxins from the body.

Days after the product had been launched, it faced official censure. The UK Advertising Standards Authority confirmed that the advertising for the tincture violated UK advertising rules.[184] In addition, scientists, including myself, publicly criticised the product for several other reasons. I pointed out that the sale of the Detox Tincture raised several important points:

- The pathophysiology of 'detox' flies in the face of science.
- As a therapeutic approach, detox is implausible, unproven, and dangerous.
- Charles ignored science and preferred to rely on 'make believe' and superstition.

I also argued that the promotion of bogus detox products may contribute to ill health by suggesting we can happily over-indulge, then take a detox remedy and be perfectly fine again. Under the banner of holistic and integrative healthcare, I suggested, Charles promoted a 'quick fix' and outright quackery. Others criticised Duchy Originals for claiming that the tinctures had been tested for efficacy, and for basing this claim on the approval by the UK regulator, the MHRA. However, the MHRA specifically stated that they do not test medicines for efficacy and that they were only tested for safety and quality.

After Charles' detox adventure had been discussed extensively in the press for several days, Duchy Originals changed the product description, and Duchy Originals felt the need to put the following statement up on their blog:

> Following recent press articles regarding our Duchy Herbals range, we are aware that some of our customers may be seeking reassurance about the range. Our CEO, Andrew Baker, says: 'Together with our partners, Nelsons—leaders in the field of natural medicine, we spent many years researching and developing our first range of herbal tinctures. It is a range that we are truly proud of. Our Duchy Herbals Echina-Relief Tincture and Duchy Herbals Hyperi-lift Tincture have both been approved and licensed as traditional herbal medicines by the UK regulatory authorities, the Medicines and Healthcare Products Regulatory Agency (MHRA). Our Duchy Herbals Detox Tincture, which is traded as a food supplement, has been produced to the highest quality standards and within the

regulatory framework of both UK and European food law. I do hope, therefore, that you are able to share our confidence in the compliance of the Duchy Herbals range to the very highest regulatory standards.' We'd love to hear what you think, click on comments below and send us your thoughts.[185]

21.3. Consequences

Amidst international criticism combined with a good deal of ridicule, the tincture was eventually withdrawn from the market. Charles' foray into commercial herbalism was thus not merely a financial failure, but also turned out to be a further reputational damage. The prominent US neurologist and blogger, Steven Novella, for instance, wrote:

> If I tried to invent a product name that evoked the sense of patent-medicine snake oil from the 19th century I don't think I could have done a better job. Duchy Originals detox tincture is just one more of thousands of snake oil products being marketed to the public with dubious health claims. Except this one is backed by the Prince of Wales.
>
> The con is an old one — virtually random ingredients are put into a pill, elixir, tincture, or salve and sold with incredible hype but no science. So-called snake oil marketers have a long tradition of knowing their marks and the market. Claims are designed to appeal to the broadest market, to have maximal allure, and to be just vague enough to evade any pesky regulations that may be in effect. Claims also tend to follow recent fads, using the buzz-words that are hot, and often try to wrap cutting-edge sciency terms in the cloak of ancient wisdom...[186]

And the British Dietetic Association stated that 'the idea of "detox" is a load of nonsense', adding that 'there are no pills or specific drinks, patches or lotions that can do a magic job... for the vast majority of people, a sensible diet and regular physical activity really are the only ways to properly protect your health'.[187]

Many others were much less respectful, e.g. 'Charlie provides no evidence whatsoever that drops of artichoke and dandelion will detox your system, whatever that means. So I'm left to assume that this is just another of his money making schemes. The trouble is of course that with his high profile and royal connections an awful lot of people out there will fall for it and pass over their hard earned cash.'[188]

Box 21

Examples of other alternative therapies promoted as detox treatments

- Ayurvedic medicine
- Chelation therapy
- Coffee enemas
- Colonic irrigation
- Cupping
- Diets
- Ear candles
- Foot baths
- Gerson therapy
- Gua Sha
- Herbal medicines
- Homeopathy
- Homotoxicology
- Leech therapy
- Narconon programme (Scientology)
- Naturopathy
- Supplements

Twenty-Two

Charles' Letters to Health Politicians

Charles' role as the heir to the throne did not include interfering in politics. Yet, he did write to UK politicians on a regular basis, not just in relation to health politics. In 2015, after a lengthy and complicated legal battle,[189] some of his letters written between September 2004 and March 2005 were disclosed to the public. They were nicknamed the 'black spider memos' because of King Charles' messy scrawl in black ink (Box 22).

22.1. Charles' views

After the memos were released, the palace issued a statement expressing dismay that the letters were made public: 'The publication of private letters can only inhibit his [King Charles'] ability to express the concerns and suggestions which have been put to him in the course of his travels and meetings.' The statement furthermore asserts that the then-prince 'cares deeply about this country, and tries to use his unique position to help others'.

The memos reveal that Charles repeatedly wrote to politicians attempting to protect or promote alternative medicine (Chapter 15) or demanding greater access to alternative therapies in the NHS alongside conventional medicine.[116] For instance, in 2007, he wrote to the then Health Secretary, Alan Johnson, saying that, 'despite waves of invective over the years from parts of the medical and scientific establishment', he continued to lobby 'because [he] cannot bear people suffering unnecessarily when a complementary approach could make a real difference'.

He also opposed 'large and threatened cuts' in the funding of homeopathic hospitals and their possible closure and stated that referrals to the Royal London homeopathic hospital were increasing 'until what seems to amount to a recent "anti-homeopathic campaign"'. He warned against cuts and claims 'that these homeopathic hospitals deal with many patients with real health problems who otherwise would require treatment elsewhere, often at greater expense'.[189]

Charles furthermore suggested that illness should be treated with a 'whole person approach' rather than a 'reductionist focus on the particular ailment'. And he urged that alternative remedies should be used to treat 'effectiveness gaps' in mainstream medicine, especially with musculoskeletal problems, depression, eczema, chronic pain, and irritable bowel syndrome.[190]

22.2. The evidence

Charles' memos related to alternative medicine contain several statements that deserve a critical comment. Charles' statement that he 'cannot bear people suffering unnecessarily when a complementary approach could make a real difference' seems particularly odd. Does he see himself in a better position than doctors to estimate the suffering of patients? The truth is that he has very little contact with the average NHS patient; his circle of friends and acquaintances are likely to be able to afford private healthcare whenever they need it and can easily chose any alternative therapies they want. Crucially, he is not educated or trained to competently judge whether alternative treatments can make a 'real difference'. At the time of Charles' statements, my team had published easily accessible summaries of the evidence which clearly showed that only very few alternative therapies might ease the suffering of patients who are severely ill.[191]

What Charles views as an 'anti-homeopathic campaign' was, in fact, a drive towards evidence-based medicine in the best interests of the patient. If patients are truly suffering, they obviously need the most effective treatments available. If Charles believes that homeopathic remedies fall into this category, he is quite simply wrong.[192] Not offering the best possible treatment is not in the best interests of patients and is deeply unethical.[193]

Charles' assumption that homeopathy might treat patients 'with real health problems' at lower cost than conventional medicine is erroneous too (Chapter 16). If a therapy is not effective, it cannot adequately deal with the problem at hand. Consequently, the health problem will need to be addressed by effective means at a later stage. To put it simply: there cannot be cost-effectiveness without effectiveness. This means that homeopathy cannot save money, but in fact incurs extra costs.[194]

Charles' assumption that alternative therapies are convincingly effective in the treatment of musculoskeletal problems, depression, eczema, chronic pain, or irritable bowel syndrome is based almost entirely on Charles' wishful thinking.[190] Someone with close links to the Palace said: 'It is no accident that he writes all those letters to ministers. He [Charles] does see himself as a kind of saviour of the nation, someone who can mend the broken country. Some might see that as presumptuously messianic.'[195]

What is perhaps most important about Charles' letters and his many other attempts to force alternative medicine on to the NHS is the fact that they are attempts to meddle in health politics. We have seen in previous chapters that, in his enthusiasm for alternative medicine, Charles seems to overstep this mark regularly. Yet many experts believe that, as the heir to the thrown, he would be well advised to stay out of politics; a notion that seems to be in agreement even with what Charles published on his own official websites:

> The main part of The Prince of Wales's role as Heir to The Throne is to support Her Majesty The Queen as the focal point for national pride, unity and allegiance and bringing people together across all sections of society, representing stability and continuity, highlighting achievement, and emphasising the importance of service and the voluntary sector by encouragement and example.[196]

22.3. Consequences

After the disclosure of the memos, restrictions have been tightened. It is therefore unlikely that royal correspondence of this nature will be released again. Yet, as a result of the repeated interferences in UK politics, Charles has often been referred to as 'the meddling Prince'.

Despite these incessant activities, it seems fair to say that, in the realm of alternative medicine, Charles' interventions had little impact. If anything, the availability of alternative therapies on the NHS has since decreased. Today, few politicians seem to take much notice of his promotion in alternative medicine. They would, of course, listen politely but then they simply seem to ignore it.

The Birmingham Labour MP Steve McCabe said it was 'strange' that the heir to the throne should be able to lobby the Health Secretary on such a controversial issue. 'It is even more extraordinary that he should be allowed to do this in secret... I can't see how it isn't in the public interest for the rest of us to know', he said. His colleague, Paul Flynn, claimed Charles had a duty to remain neutral, particularly over a hugely controversial issue involving public spending and the health of the nation: 'People are entitled to believe what they want, but having the heir to the throne attempting to influence the spending of precious NHS resources on a service he probably doesn't use at all is ludicrous... Prince Charles should not be interfering; he is in training for his role as monarch and the first lesson is to put a bandage round his mouth and to keep it there at all times.'[197]

Box 22

The 'black spider memos'

- In 2010, a *Guardian* journalist made an application under the 'Freedom of Information' law for the release of Charles' letters to politicians.
- Permission was initially refused, and a lengthy legal battle ensued.
- In 2015, the court ordered much of Charles' correspondence with UK government ministers to be released.
- The subjects addressed in these letters include alternative medicine, farming, genetic modification, global warming, social deprivation, planning, and architecture.
- An editorial in *The Guardian* stated that 'the letters show that behind that curtain, most of the time, Prince Charles behaves more as a bit of a bore on behalf of his good causes than as any sort of wannabe feudal tyrant'.[198]

Twenty-Three

The College of Medicine and Integrated Health

Following its ignominious closure, many of the individuals and institutions that had been involved in Charles' Foundation for Integrated Health (Chapter 10) launched a new organisation (Box 23). In October 2010, only weeks after the foundation's demise, they founded 'The College of Medicine', which was later renamed 'The College of Medicine and Integrated Health'. The chair, Dr Michael Dixon (Box 10), had previously been the medical director of the Foundation for Integrated Health. The mission of the college as currently posted on their website is as follows:

> The College of Medicine was founded in 2010 to reform healthcare so that it works for everyone in a way that's inclusive, progressive and compassionate. We want to redefine medicine beyond pills and procedures, to reconnect practitioners with patients, people with their environment and use both conventional and non-conventional approaches to health. The College is a 'coming together' of some of the brightest minds in the UK including NHS pioneers, scientists, CAM professionals, students and members of the public.[199]

23.1. Charles' involvement

To start with, Charles was not officially involved in the college. He did, however, address their conferences and was unofficially engaged in many of the college's activities. To most observers, it was clear that the college was the successor of the foundation and that Charles, after a period of keeping a low profile, would in some shape or form join. In 2019, Charles became the patron of the College. Dr Dixon commented: 'This is a great honour and will support us as an organisation committed to taking medicine beyond drugs and procedures. This generous royal endorsement will enable us to be ever more ambitious in our mission to achieve a more compassionate and sustainable health service.'[200]

23.2. The evidence

Several commentators confirmed that the new college is simply a rebranding of the Prince's Foundation of Integrated Health.[201] Prof David Colquhoun noted that the college is extremely well-funded and, from its beginning, seemed very confident of the Prince's support; explicitly describing its mission as 'to take forward the vision of HRH the Prince of Wales'.[202]

In a 2013 interview, Dixon was asked: 'Will alternative medicine be taught in UK universities?' He answered: 'The College of Medicine, UK, is fighting hard for it. We are historically drenched in conventional medicine and to think out of the box will take time. But we are at it and hope to have it soon.'[203] This statement makes the true mission of the college much clearer: it foremost seems to aim to champion alternative medicine. Even Wikipedia states bluntly: 'it [the college] promotes alternative medicine.'[204] And the college itself states that its aim is 'promoting, fostering and advancing an integrated approach to health care'.[205] In the pursuit of this aim, the college gets involved in a multitude of activities. In 2016, for instance, they announced the 'Integrated Medicine Diploma Course', 'the only accredited Integrative Medicine diploma currently available in the UK… [It] will provide you with an accredited qualification as an integrative medicine practitioner. The Diploma is certified by Crossfields Institute and supported by the College of Medicine and is the only one currently available in the UK'.[206]

The Crossfields Institute promotes Rudolf Steiner's anthroposophy[207] and is chaired by Simon Fielding, OBE, an osteopath and a trustee of the College who was the first chair of the General Osteopathic Council, responsible for bringing the Osteopaths Act into force in 1993 (Chapter 8). He has also served as a long-term trustee on the boards of the Foundation for Integrated Health and as the founder chair of the Council for Anthroposophical Health and Social Care'.[208]

The College also offers training in all sorts of other bizarre and non-evidence-based treatments, such as neurolinguistic programming (NLP). This treatment was developed in the 1970s and is not easy to define.[209] Even NLP experts use such vague language that can mean different things to different people. One claim, for instance, is that NLP helps people to change by teaching them to program their brains. We were given brains, we are told, without an instruction manual, and NLP offers a user's manual for the brain. The college started advertising courses on NLP for GPs and alternative medicine practitioners in 2020 with this announcement:

> Neurolinguistic Healthcare in association with the College of Medicine brings you a 2-day Introduction to Hypnosis, Neurolinguistic Programming (NLP) and Neurolinguistic Healthcare (NLH). Dr Wong and Dr Akhtar who lead the course are Trainers in NLP and Hypnosis and GPs who apply their skills in daily practice within the 10-minute consultation. The course is suitable for both medical professionals and complementary therapists. This is a limited training event offered by them to share their years of knowledge and skills with you.

The evidence for NLP is not encouraging; a systematic review concluded that 'there is little evidence that NLP interventions improve health-related outcomes.'[210]

An even more exotic treatment is thought field therapy (TFT). The college advertised it with the following words:

> As part of our ongoing programme to explore prospects for improved healthcare, the College is pleased to announce a course on TFT—a 'Tapping' therapy—independently provided by Janet Thomson MSc.

> In healthcare we may find ourselves exhausting the evidence-based options and still looking for ways to help our patients. So when trusted practitioners suggest simple and safe approaches that appear to have benefit we are interested.
>
> TFT is a simple, non-invasive, technique that anyone can learn, for themselves or to pass on to their patients, to help cope with negative thoughts and emotions. It was developed by Roger Callahan who discovered that tapping on certain meridian points could help counter negative emotions. Janet trained with Roger and has become an accomplished exponent of the technique.
>
> Janet has contracted her usual two-day course into one: to get the most from this will require access to her Tapping For Life book and there will be pre-course videos demonstrating some of the key techniques. The second consecutive day is available for advanced TFT training, to help in dealing with difficult cases, as well as how to integrate TFT with other modalities.[211]

Thought field therapy is a fringe psychological treatment that is claimed heals a variety of mental and physical ailments through specialised 'tapping' with the fingers at meridian points on the upper body and hands. The theory behind TFT is a mixture of concepts 'derived from a variety of sources. Foremost among these is the ancient Chinese philosophy of chi, which is thought to be the "life force" that flows throughout the body'. There is no scientific evidence that TFT is effective, and the American Psychological Association has stated that it 'lacks a scientific basis' and consists of pseudoscience.[212]

The colleges' 'Self Care Tool Kit' is an initiative that started in 2009 with a national multi-centre project commissioned by the UK Department of Health, to consider the best way to integrate self-care into family practice.[213] The project involved two large family health centres and two university departments. One output was the Self Care Library (SCL), an online patient resource providing free information about self-care. When the funding for the SCL ceased, the facility was assigned to the college. With funding from 'Pukka Herbs Vitamins, Herbal Remedies & Health Supplements', the College was able to transfer the SCL content. Simon Mills, the coordinator of the original project who is also an employee of Pukka, has led this transformation and helped the College set up the new parent portal, Our Health Directory.[214] The Self Care Toolkit is thus a revamped version of the SCL. This begs the question whether it is wise to have a commercial sponsor for such a project. The entry on depression, for example, contains phraseology that is likely to encourage patients suffering from this condition to buy the supplements mentioned:

- '[P]eople with low blood levels of the B vitamin folic acid are more likely to be depressed and less likely to do well on anti-depressant medicines…'
- 'Some people say that taking high doses of vitamin C (1–2g and more a day) helps lift their mood…'
- 'There is a little research to support this and none showing that high doses of vitamin C actually help clinical depression…'
- 'A lack of vitamin D can lead to depression.'
- 'Taking supplements of vitamins B and D might help some people…'
- '…[Y]our GP will prescribe it for you or you can buy a vitamin D supplement.'

The evidence does not support any of the mentioned supplements —all sold by Pukka—as a treatment for depression.[215,216,217] Furthermore, over-dosing some of these vitamins can cause serious health problems. Finally, one should mention that depression is, of course, a life-threatening illness. There is thus a danger that patients who trust the College's pronouncements might commit suicide because of the ineffectiveness of these therapies.

On the occasion of celebrating the college's tenth anniversary in 2020, Sir Cyril Chantler praised the college for its focus on complementary medicine and urged more support for research, saying: 'The role of the College of Medicine is to promote complementary medicine and holistic medicine and the promotion of health.' He added: 'I have no doubt whatsoever that if practised responsibly, there is plenty of evidence of benefit to patients. The

College is important, not only to promote, but also to police practice and to ensure as far as possible that safety is a prime concern.'[218] The mention of safety seems important. Yet, as far as I can see, none of the College's activities is focused on assessing the risks of alternative medicine.

23.3. Consequences

Today, the college pursues multiple activities and is much admired by enthusiasts of alternative medicine. Outside this realm, however, it has so far failed to have a significant impact. On the contrary, the college has attracted much criticism from a diverse range of sources, as quoted above.

Possibly realising that the concepts of integrated medicine are deeply flawed (Chapter 13) and therefore of limited long-term potential, the college has recently started to focus on 'social prescribing' (Chapter 27).

Box 23

Prominent proponents of alternative medicine within the College of Medicine and Integrated Health

- Akhtar, Naveed (GP, NLP trainer, hypnotherapist, acupuncturist)
- Dixon, Michael (GP) chair, trustee and council member
- Fielding, Simon (osteopath) trustee
- Mills, Simon (herbalist) council member
- Ornish, Dean (doctor) college ambassador and international ambassador
- Peters, David (GP, osteopath) council member and member of the scientific advisory council
- Robinson, Nicola (Professor of Traditional Chinese Medicine and Integrated Health) member of the scientific advisory council
- Sikora, Karol (oncologist) member of the scientific advisory council
- Thompson, Edward (GP, homeopath) council member
- Thompson, Elizabeth (homeopath) council member
- Weil, Andrew (integrative medicine) international ambassador
- Zollman, Catherine (GP) council member

Twenty-Four

The Enemy of Enlightenment

The Enlightenment was a philosophical movement that started in the late seventeenth century. It emphasised reason and individualism rather than tradition and was greatly influenced by eminent thinkers such as Descartes, Locke, Newton, Kant, Goethe, Voltaire, Rousseau, Spinoza, Leibnitz, and Adam Smith. The 'age of reason', as the Enlightenment is also often called, stressed the importance of science and evidence over intuition and religion as the main source of knowledge. Today, the Enlightenment is celebrated as a symbol for rational thinking and scientific analysis.

24.1. Charles' views

Charles has made numerous statements that could have made rational thinkers suspect he was not just doubtful of reason and science but intuitively poised against them. In 2010, this attitude was highlighted by his admission to being proud of being called the enemy of the Enlightenment: 'I was accused once of being the enemy of the Enlightenment... I felt proud of that.'[219,220]

24.2. The evidence

With this statement, Charles placed himself firmly into the tradition of the Romantics, who felt that 'the age of reason' neglected qualities such as imagination, intuition, and the mysteries of the soul. Prof David Colquhoun, FRS, commented on Charles' statement: 'That's a remarkable point of view for someone who, if he succeeds, will become the patron of that product of the age of enlightenment, the Royal Society. I have no doubt that Prince Charles means well. He can't be blamed for his lack of education. But his views on medicine date from a few centuries ago, and he has lost no opportunity to exploit his privileged position to proclaim them.'[221]

The Enlightenment was the period in which scientific empiricism and rational thought first came to the fore. It was the time when scientific societies and academies became centres of scientific research and development. Many important advancements in medicine and other fields were achieved during this period. By claiming to be proud of being called the enemy of the Enlightenment, Charles confirmed his confused attitude towards science, which had previously shone through in many of the statements already mentioned in previous chapters, for example:

- '...[O]ur outlook in the Westernized world has become far too firmly framed by a mechanistic approach to science...' (Chapter 4).
- '...[I]t is God's will that the unorthodox individual is doomed to years of frustration, ridicule and failure in order to act out his role in the scheme of things, until his day arrives and mankind is ready to receive his message...' (Chapter 6).
- 'The good doctor's therapeutic success largely depends on his ability to inspire the patient with confidence and to mobilise his will to health...' (Chapter 6).
- 'Scientific progress comes as much through deductive logic, rational debate and critical evaluation as it does through intuitive reasoning, creative play and the ability to tolerate uncertainty...' (Chapter 7).
- 'Recognizing the state of disharmony in the body, they [osteopaths] work to restore balance towards a state of harmony and health' (Chapter 8).
- '...[O]steopathy offers an integrated and preventive system of healthcare...' (Chapter 8).
- '...[W]e should be mindful that clinically controlled trials alone are not the only prerequisites to apply a healthcare

intervention. Consumer-based surveys can explore why people choose complementary and alternative medicine and tease out the therapeutic powers of belief and trust' (Chapter 11).

'There are different types of evidence, and the evidence of experience is just as important as scientific evidence', Charles was once reported as saying.[20] This notion might be the key to understanding his attitude towards alternative medicine. In just a few words, it encapsulates his dislike for scientific empiricism and his disdain for the role 'scientific evidence' has played in healthcare during the last 150 years.

Experience in healthcare is undoubtedly important, but it is fundamentally different from evidence (Box 1). Whenever a patient experiences a symptomatic improvement after receiving a treatment, it is assumed that this outcome must be caused by the treatment. However, this assumption ignores several fundamental facts. Two events — the treatment and the improvement — that follow each other in time are not necessarily causally related (to argue otherwise means employing the infamous 'post hoc ergo propter hoc' fallacy[222]). There are several factors that determine a clinical improvement which are unrelated to the treatment itself:

- The natural history of the condition: many conditions get better by themselves regardless of whether we treat them or not.
- Regression towards the mean: extremes tend to be less extreme when they are measured a second time.
- The placebo effect: conscious expectation and subconscious conditioning can influence the outcome without any contribution of the therapy *per se*.
- Concomitant treatments: most patients use several therapies in parallel which makes it impossible to determine which caused the improvement.

These phenomena determine the clinical outcome in such a way that inefficacious therapies, or even mildly harmful treatments, can appear to be efficacious. In other words, the prescribed treatment is only one of the determinants of the clinical outcome, and even the most impressive clinical experience can be (and often is) totally misleading. To put it bluntly: Charles' statement about the evidence of experience is quite simply wrong. Experience is essential in healthcare, but it never amounts to reliable evidence.[223]

Such considerations are by no means revelations that are accessible only to the anointed; they are discussed in virtually every text about medical evidence. Charles is evidently not aware of them — or is he? Charles' lack of enlightenment might just be due to motivated ignorance. Ignorance is, of course, the lack of knowledge; and motivated ignorance is the lack of knowledge due to a conscious decision to ignore information that, for some reason or other, is perceived as inconvenient, disturbing, or unwelcome.[224] Motivated ignorance is a well-studied phenomenon. It is a means of avoiding disconcerting information and gives proponents of alternative medicine (or any other creed) the opportunity to shut their eyes and ears to the uncomfortable reality that their favourite treatment might be not as fabulous as they think. The phenomenon is especially powerful when someone has acquired a quasi-religious belief in alternative medicine. It is hard to convince religious believers that their god does not exist. It is equally difficult to convince believers in alternative medicine that their favourite alternative therapy is bogus. For people like Charles who are bent on rejecting science and reason, even the most persuasive evidence will have little impact.

A possibility to minimise motivated ignorance would be to learn the skill of critical thinking. This would involve reading and discussing views that differ from one's own. It might also mean listening to advice from leading experts in the field of alternative medicine. And it certainly involves asking oneself whether a dearly held belief is correct. Sadly, this seems not what Charles wants or is able to consider, practise, or tolerate. By admitting to being a proud enemy of the Enlightenment, Charles effectively outed himself as the champion of the 'age of unreason'. In a *Lancet* editorial about alternative medicine and the 'age of unreason', Petr Skrabanek once put it nicely:

> …[W]hat is at issue is the complex problem of demarcation between science and quackery, between reason and faith,

between honest search for truth and unscrupulous exploitation of human suffering... Inventors of perpetual-motion machines cannot reasonably expect that judgment on their claims should be suspended until they are proved false by the non-believers. Outrageous claims are often ignored by sceptics, who may be justified in their sense of the futility of it all. 'It is like punching a feather pillow—an indentation is made, but soon refills, and the whole soft, spongy mass continues as before.' The stoic silence of sceptics is then interpreted by irrationalists as lack of argument; the confused public, always inclined to be deceived, gives the irrationalists the benefit of the doubt.[225]

The often confused and royalty-admiring public does indeed tend to give Charles the benefit of the doubt. But many critical thinkers have lost patience with his incessant irrationality:

> His support for the environment, Georgian architecture and organic biscuits make him seem a 'conservative' in the broadest sense... To call Prince Charles a conservative thinker is like calling a flat-earther a conservative geographer. He most admires primitive societies dominated by shaman kings—who look rather like him—and inevitably regards knowledge as suspect.[226]

24.3. Consequences

Charles' announcement that he is proud to be called an enemy of the Enlightenment was widely reported. For rational thinkers, it provided the last proof required for characterising him as profoundly anti-scientific. Some even saw it as a declaration of war on reason itself.

However, in the realm of alternative medicine, the 'age of unreason' has become, not least thanks to Charles, a lamentable fact. Proponents of alternative medicine feel today entitled to apply double standards—reason for medicine and unreason for alternative medicine. This attitude can be seen in numerous documents, but perhaps nowhere clearer than in the unwieldy paper entitled 'Integrated Healthcare: A Way Forward for the Next Five Years? A Discussion Document from the Prince of Wales's Initiative on Integrated Medicine'.[227] It is a perfect example of special pleading for alternative medicine. For instance, it discusses about 30 different research methodologies for investigating alternative medicine, while stressing that 'it is by no means the case that randomised clinical trials (RCTs) will always be the appropriate choice'. It is true, studies to determine whether alternative treatments work (i.e. RCTs) are not the only way to do research. But without knowing whether homeopathy, for example, works, it is futile to investigate other questions such as: how we can improve the referral systems to homeopaths? Can homeopathy be understood within a common framework? Or any of the other research questions suggested in this document.

In the world of science, Charles' proclamation of the dawn of 'the age of unreason' might have merely provoked ridicule. In alternative medicine, however, it had a more detrimental impact. Essentially, it entitled proponents of even the most bizarre concepts to expect to be taken seriously. This, in turn, led to a situation where, instead of focusing on investigating the safety and efficacy of the few promising modalities, efforts and funds were wasted on hopelessly implausible treatments and ideas. In the end, this has helped nobody, least of all alternative medicine itself.

Box 24

Science, pseudoscience, and anti-science

- Science is the identification, description, observation, experimental investigation, and theoretical explanation of phenomena.
- Pseudoscience is anything that tries to imitate science but does not meet its standards nor abide by its rules.
- Pseudoscience is dangerous because most people find it difficult to tell it from science and are thus misled by it into making wrong decisions.
- Anti-science is a set of attitudes that involve the outright rejection of science and the scientific method.
- People holding anti-scientific views do not accept science as a method for generating universal knowledge.

Twenty-Five

Harmony

In 2010, Charles published his book entitled *Harmony: A New Way of Looking at Our World*. It was Charles' first attempt to explain his thinking in full detail and to join up all his ideas under one all-encompassing umbrella. This renders the analysis of *Harmony* a central issue and makes the present chapter a focal point of this biography (Box 25).

For this book, Charles had recruited two co-authors, Tony Juniper[228] and Ian Skelly.[229] He also acknowledged the help of two experts in alternative medicine, Michael Dixon (Chapter 23) and Mosaraf Ali (Chapter 12). This is relevant not least because such assistance should assure that the technical details related to alternative medicine were checked and are factually correct.

The book is very well presented and contains a large number of high-quality photographs. Unsurprisingly, it contains several sections that are relevant to the subject of alternative medicine.

25.1. Charles' intention

Charles' book was intended as a

> practical guide to what we have lost in the modern world, why we have lost it and how easily it is to rediscover. *Harmony* is a blueprint for a more balanced, sustainable world that the human race must create to survive… In *Harmony*, Prince Charles looks at different aspects of our modern world to demonstrate how many of the challenges seen in areas as diverse as architecture, farming and medicine can be traced to how we have abandoned a classical sense of balance and

proportion. From the rice farms of India to America's corn belt, *Harmony* spans the globe, dissecting the specific practices of modern life that have put us at odds with the world and showing how this imbalance manifests itself throughout our lives. *Harmony* shows how the imbalance that has emerged is at heart of a crisis which now threatens our very civilisation. It tells the story of how our disconnection from Nature has contributed to the greatest crisis in the history of mankind and how seeking balance in our actions will return us to a more considered, secure, comfortable and cleaner world. Drawing on his own practical experience, Prince Charles charts how changes to how we look at the world could lead us toward a better future. He describes how knowledge and perspectives now largely lost could help us meet very modern challenges, including in the built environment, engineering, medicine and farming.[230]

25.2. The evidence

Charles' book was well received by many of its readers. One of the most enthusiastic reviews on Amazon, for instance, stated:

> This book represents the finest and clearest distillations of wisdom that can be found in any book or any previous individual mind. Harmony is creating and maintaining simple harmonic relationships between our thoughts, speech, and actions, which is our True Nature. Yet, how extraordinary that it has taken 800 years of breeding to produce such an important messenger, and the most fitting, next king and great teacher.[230]

Other readers were more reserved or even outright sceptical:

> 'Harmony' is a collection of New Age scripture ghosted for Prince Charles by one of the Friends of the Earth. The offering is very much about the Prince and he keeps popping up like a rather grand Forrest Gump beside Buddhist temples and albatross colonies. At times it becomes very bizarre and the future head of the Church of England links Osiris and Jesus in

a way that would once have had him burnt at the stake. He also presents his weird views as important personal revelations but dismisses opposing views from medics or scientists as cynical blindness. It includes all his usual homeopathic quackery, his insistence that GM crops are 'the devil's work' and such global-warming tosh as would embarrass even Al Gore. It is an interesting insight into his mind but the book itself is not much more than the wish list of a Buddhist mystic.[230]

Even though the book was accompanied with a promotional film and was launched with a maximum PR effort, it only sold fewer than 15,000 copies (of which 2,000 were free hand-outs at the film premiere).[231] Yet, Tony Juniper refused to accept failure and assured Charles that 'it will take ten years before we know the impact of Harmony'.[229] Nearly a decade later, Nick Cohen did not seem impressed by the book's impact and commented in the *Spectator*:

> As the few people who have read his book *Harmony* know, his world view isn't so much old-fashioned as pre-scientific. It is from his combination of mysticism and obscurantism that the Prince's otherwise startling denunciation of Galileo for introducing 'mechanistic' thought into a previously pure world flows. Hence his support for every variety of 'alternative' quackery and pseudo-science, and his hope that we will 'grow numb at the sacred presence all traditional societies feel'.[232]

In many passages, the book confirms Charles' rejection of science.[233] Here are just four examples:

- 'Essentially it is the spiritual dimension to our existence that has been dangerously neglected during the modern era—the dimension which is related to our intuitive feelings about things.'
- 'The language of empiricism is now so much in the ascendant that it has authority over any other way of looking at the world. It decides whether those other ways

of looking at things stand up to its tests and therefore whether they are right or wrong.'
- 'Even many people in the West fail to recognize that so much modern science is not simply an "objective" knowledge of Nature, but is based upon a particular way of thinking about existence and geared to the ambition to gain dominion over Nature. The way in which this has happened has a lot to do with the numbing of our vital inborn or "inner tutor", the so-called human "intuition".'
- '...scientific rationalism continues to turn people away from any form of spiritual practice or reflection by perpetuating what seems to me to be a widespread confusion.'

In the context of alternative medicine, a few passages deserve closer scrutiny. Here are a few points that might be relevant to our analysis of Charles' thinking, all from the section entitled 'Complements and alternatives':

- Charles states that 'rivers flow just as our blood flows, by virtue of spirals'. Having completed my PhD in blood rheology, I can assure Charles and everyone else that blood does not flow in our bodies by virtue of spirals.
- '...[O]ur outlook is predominantly mechanical, and brings us back to the shortcomings of our mechanistic view of the world.' This might be so for osteopaths who, as we shall see, Charles praises enthusiastically, but it certainly does not hold true for conventional medicine. For instance, pharmacology, the field Charles feels is far too dominant in modern medicine, can hardly be called mechanistic.
- '...[T]here is a great deal to be gained from complementary treatments... traditional methods can deliver huge benefits...' This is what Charles believes and what he has been repeating for decades. However, he rarely provides any convincing evidence in support of such assumptions—mostly, one must presume, because there hardly is any. Thus these statements can easily be disclosed as examples of wishful thinking.

- '...I cannot bear to see people suffer unnecessarily when, so often, a complementary treatment can be beneficial...' This is another theme that Charles likes to stress. It is wholly unjustified and could even be perceived as insulting to those who actually do see patients who suffer and try their best to help them by employing the most effective treatments available (which only rarely would include alternative therapies).

While these points could perhaps be viewed as somewhat trivial, Charles' discussions of several specific alternative therapies are certainly not. In his book, he elaborates on osteopathy, which we have dealt with in Chapter 8, on homeopathy, which we will discuss in Chapter 29, and on Ayurvedic medicine, which is the focus of Chapter 27.

Of acupuncture, a popular alternative therapy that Charles had previously supported on a regular basis (for instance by addressing conferences of UK acupuncture organisations), Charles simply claims that it is a 'proven discipline' (p. 321). This statement is, however, far from borne out by the evidence. In 2009, the year when *Harmony* was presumably being written, I published an overview of the most reliable evidence. Here is its summary:

> Many trials of acupuncture and numerous systematic reviews have recently become available. Their conclusions are far from uniform. In an attempt to find the most reliable type of evidence, this article provides an overview of Cochrane reviews of acupuncture. Such reviews were studied, their details extracted, and they were categorized as: reviews with a negative conclusion (no evidence that acupuncture is effective); reviews that were inconclusive; and reviews with a positive or tentatively positive conclusion. Thirty-two reviews were found, covering a wide range of conditions. Twenty-five of them failed to demonstrate the effectiveness of acupuncture. Five reviews arrived at positive or tentatively positive conclusions and two were inconclusive. The conditions that are most solidly backed up by evidence are chemotherapy-induced

nausea/vomiting, postoperative nausea/vomiting, and idiopathic headache. It is concluded that Cochrane reviews of acupuncture do not suggest that this treatment is effective for a wide range of conditions.[234]

Charles also mentions 'traditional methods of diagnosis'. We often consider alternative medicine to be a conglomerate of diverse therapies and tend to forget that it also includes a diverse array of diagnostic techniques.[29] Specifically, Charles tells us that iridology, as well as tongue and pulse diagnostic techniques and foot reflexology, are based on 'wisdom accumulated over thousands of years' and are valuable additions to modern diagnostics. 'Why persist in denying the immense value of such accumulated wisdom when it can tell us so much about the whole person — mind, body and spirit?'

Charles' question is best answered by summarising the evidence on the reliability of these techniques. For all of the diagnostic methods which Charles mentions, scientific tests have been conducted and are publicly available for anyone who cares to look for them.

Iridology is a diagnostic method based on the belief that pigmentations on specific spots of the iris of patients provide diagnostic clues as to the health of their organs. Iridologists believe that the iris is a 'mirror of our body'. Any relevant abnormality on the right half of the body will reveal itself on the right iris and problems on the left side will show up on the left iris. They assume that the iris is linked via multiple nerve connections to all organs and believe that any bodily malfunction will thus be represented as abnormalities of pigmentation on the iris. These assumptions are not in keeping with basic anatomy or physiology and thus lack plausibility. Iridologists have produced detailed maps of the iris where each iris is divided in 60 sectors and each segment is related to an inner organ or bodily function. Iridologists either study the iris in situ or they produce high-quality colour photographs of both irides for detailed inspection. Several studies have tested the validity of iridology. A systematic review of these data concluded that 'the validity of iridology as a diagnostic tool is not supported by scientific evaluations. Patients and therapists should be discouraged from using this method'.[235] Ophthalmologists have repeatedly warned of the use iridology.[236]

Tongue diagnosis is a method for identifying disease used in Ayurvedic and Traditional Chinese Medicine (TCM). It is different from the diagnostic clues which conventional clinicians might obtain from examining their patients' tongues. TCM practitioners look at the tongue of patients to evaluate its qualities such as colour, coating, shape, presence or absence of cracks. They also use maps where specific areas of the tongue are supposed to correspond to specific organs of the body. From this information, practitioners claim to glean diagnostic clues and guidance as to what the best therapy might be. The assumptions underlying tongue diagnosis are not in line with our current knowledge of anatomy, physiology, pathophysiology, etc. In other words, they are not plausible. Crucially, several studies have failed to produce good evidence that tongue diagnosis can reliably identify any disease or condition.[237,238,239]

Pulse diagnosis is the technique used in Traditional Chinese Medicine (TCM) and several other Asian traditions to identify the cause of ill health by feeling the pulse of a patient. Conventional clinicians feel the pulse to check the regularity and frequency of the heartbeat. This conventional form of pulse-taking is unrelated to the traditional pulse diagnosis to which Charles refers. Practitioners of TCM aim to feel other qualities of the pulse and claim they provide information about a patient's state of health. In Ayurvedic medicine, the examination of the pulse is used for identifying energy imbalances of *prana, tejas,* and *ojas*. Pulse diagnosis is based on obsolete concepts and is not plausible. In one study, the reliability of pulse diagnosis was tested by letting several experienced clinicians diagnose the same patients and comparing their findings. The authors concluded that 'the interobserver reliability in making a pulse diagnosis in stroke patients is not particularly high when objectively quantified. Additional research is needed to help reduce this lack of reliability for various portions of the pulse diagnosis'.[240]

Foot reflexology is a manual technique where pressure is applied to the sole of the patient's foot. Reflexologists have maps of the sole of the foot where the body's organs are projected. By massaging specific zones of the sole of the foot, reflexologists believe they can positively influence the function of inner organs. Reflexology is mostly used as a therapy, but some therapists also claim they can diagnose health problems through feeling tender or gritty areas on the sole of the foot which, they claim, correspond to specific organs. The assumptions made by reflexologists contradict our current knowledge of anatomy and physiology and are thus not biologically plausible. Reflexology treatment has been submitted to several clinical trials. A summary of this evidence concluded that 'the best clinical evidence does not demonstrate convincingly reflexology to be an effective treatment for any medical condition'.[241] Reflexology has also been tested for its validity as a diagnostic technique. The findings do not suggest that it is a reliable diagnostic method.[242]

These short summaries reveal that the diagnostic methods praised by Charles are, in fact, not reliable and thus are potentially harmful. The risks of unreliable diagnostic techniques consist in arriving at false-positive and false-negative diagnoses. A false-positive diagnosis would mean that a patient is told she is suffering from an illness while, in fact, she is healthy. This would cause unnecessary distress and expenses. A false-negative diagnosis means that a patient is informed that she is healthy while, in fact, she is not. This would cause delay in effective treatments and can, in extreme cases, cause the premature death of the patient.

Most of the statements on alternative medicine in Charles' book *Harmony* are thus misleading at best and dangerous at worst. Nobody, of course, would demand of the king to be fully informed on the medical evidence. Like everyone else, he has the right to be ignorant, ill-informed, or wrong. But, like everyone else, he also should either seek expert advice or remain silent about things he fails to comprehend. Charles does mention repeatedly that he got advice from 'leading experts'. I therefore fear that these advisors are perhaps not so much 'leading' as 'misleading' experts.

25.3. Consequences

With his book, Charles has provided us with a lasting document of his often bizarre views. Specifically, with regards to alternative medicine he has demonstrated himself to be ill-informed, ill-advised, and often dangerously wrong. The book's subtitle, 'A New Way of Looking at Our World', is not free of irony: his ideas related to alternative medicine are neither new (Chapter 13), nor are they addressing the realities of the world. A well-known UK sceptic commented more harshly on Charles' book:

> It would be nice to sit back and laugh at the absurd beliefs of this crank. However, Charles is no ordinary crank. He has direct access to government ministers and is prolific in his letter writing to them. He directly lobbies the Health department about these beliefs and he has the capacity to greatly reward those that comply with his wishes. And coupled with his reported dislike of criticism, he has great capacity to undermine not only public health, but the entire standing of science within government.
>
> The poet Ben Jonson said, 'They say Princes learn no art truly, but the art of horsemanship. The reason is, the brave beast is no flatterer. He will throw a prince as soon as his groom.' The Prince's ability to surround himself with toad eating flatterers means that his absurd views of medicine cannot be directly challenged. But I fear this is a Prince that needs to be thrown from the saddle of quackery, otherwise we can expect huge damage to the role of science within public life.[233]

'It will take ten years before we know the impact of *Harmony*', said Toni Juniper, one of the book's co-authors, when it was first published in 2010. Today, more than a decade later, I can attest that, in the area of alternative medicine, the impact was unimpressive, and in the field of science it was mostly negative.

Box 25

Harmony and balance

- Harmony is a pleasing combination of elements in a whole.
- Harmony is a relationship in which all components coexist without destroying each other.
- Harmony signifies a lack of conflict.
- Balance is a state of equilibrium.
- Balance is also a harmonious arrangement of all parts.

Twenty-Six

Antibiotic Overuse

Antibiotics, drugs that work against bacteria, have unquestionably saved many lives. But, if we regularly use them in situations where they are not indicated, we run into significant problems. In this case, bacteria can develop the ability to defeat the drugs designed to kill them. This phenomenon is called 'antibiotic resistance'. It means that bacteria can continue to grow unhindered and thus threaten the health of both humans and animals.

Because of antibiotic resistance, we are today often close to the most precarious point where previously treatable infections cannot any longer be defeated with antibiotics. The consequences for public health are very serious indeed.

26.1. Charles' view

Charles has repeatedly and rightly warned of antibiotic overuse. In 2016, speaking at a global leaders' summit on antimicrobial resistance, he warned that Britain faced a 'potentially disastrous scenario' because of the 'overuse and abuse' of antibiotics.

Charles then explained that he had switched to organic farming on his estates because of the growing threat from antibiotic resistance and now treats his cattle with homeopathic remedies rather than conventional medication. 'As some of you may be aware, this issue has been a long-standing and acute concern to me', he told delegates from 20 countries.

> I have enormous sympathy for those engaged in the vital task of ensuring that, as the world population continues to increase

unsustainably and travel becomes easier, antibiotics retain their availability to overcome disease… It must be incredibly frustrating to witness the fact that antibiotics have too often simply acted as a substitute for basic hygiene, or as it would seem, a way of placating a patient who has a viral infection or who actually needs little more than patience to allow a minor bacterial infection to resolve itself.[243]

26.2. The evidence

Antibiotic resistance is a very well-recognised problem, and thousands of experts worldwide are working hard to find and implement solutions. The largest databank for medical papers listed in April 2021 around quarter of a million articles on antibiotic resistance (more than 10,500 from 2016, the year of Charles' speech). Thus, Charles is not alone in worrying about its dangers to our health.

However, Charles employs a correct argument (i.e. we must use antibiotics more cautiously) to arrive at a wrong conclusion (i.e. we should all use homeopathy instead). Charles' speech therefore prompted a storm of angry reactions from scientists. Simon Singh, for instance, commented: '…Prince Charles's knee-jerk, ideological and illogical promotion of homeopathy is yet another example of him abusing his position as heir to the throne. Instead of being allowed to pontificate while being unchallenged, Prince Charles should limit himself to talking to his plants.'[244]

Charles' argument seems to be based on one of two assumptions:

- Homeopathic remedies are effective in treating or preventing bacterial infections.
- If farmers administer homeopathic remedies to their livestock, they are less likely to administer unnecessary antibiotics.

Assumption 1 can be rejected without much further debate; there is no evidence whatsoever that homeopathic remedies have antibiotic efficacy. In fact, the consensus today is that highly diluted homeopathic remedies are pure placebos (Chapter 29).

Assumption 2, however, could be seen by some as plausible and might therefore deserve further scrutiny. If we do not tell the farmers or the vets that homeopathic remedies are placebos — if, in other words, we mislead them to think they are efficacious medicines — they might use them for their animals instead of antibiotics. Consequently, the usage of antibiotics in animals would decrease. However, this strategy has obvious drawbacks:

- The truth has a high value in itself which we would disregard at our peril.
- It is not possible to keep the truth about homeopathic remedies from the farmers and we are even less able to hide it from vets.
- If we did mislead farmers and vets, we would also have to mislead the rest of the population; this would mean many patients might receive homeopathic placebos even for serious infections and other conditions which could cause many deaths of humans and animals.
- Misleading farmers and vets turns out to be unnecessary: if there is abuse of antibiotics in farming, we ought to tackle over-prescribing directly by reminding vets what they learnt in veterinary school: only use antibiotics where indicated.

Whichever way one twists and turns assumption 2, one fails to arrive at a remotely sensible conclusion. But this is not to say that Charles did not make several reasonable points in his speech. Sadly, he then spoils it all with his passion for alternative medicine.

- Yes, we have overused antibiotics both in human and in veterinary medicine.
- Yes, this has gone so far that it now endangers our health.
- Yes, it is a scandal that so little has happened in this respect, despite us knowing about the problem for decades.
- **No**, homeopathy or any other alternative medicine is not the solution to any of these problems.

If we want to reduce the use of antibiotics, we need to stop employing them in human and veterinary medicine for situations where they are not necessary. Moreover, in farming, we must improve husbandry such that antibiotics are not required for disease prevention. To a large extent this is a question of educating those who are responsible for prescribing and administering antibiotics. Education should not be as complex as it may look: healthcare professionals would merely need to abide by the principles of evidence-based medicine and by what they have been taught during their training!

26.3. Consequences

The subject of antibiotic resistance is a classic example where Charles, full of good will, starts from a reasonable vantage point but then stumbles over his own obsession with alternative medicine. This is similar to his promotion of integrated medicine where he rightly points out that modern medicine has important deficits (Box 26) and then destroys his initially reasonable argument by insisting that more use of unproven treatments is the solution (Chapter 13).

The inevitable result is that his effort to generate a positive outcome falls apart. In the end, his actions could easily turn out to be counterproductive, as his rational argument about the indiscriminate use of antibiotics is invalidated by his irrational promotion of homeopathy.

Box 26

Examples of deficits of conventional medicine and Charles' solutions to them

DEFICIT	CHARLES' PROPOSAL	COMMENT
Lack of empathy and compassion	Integration of alternative therapies	If alternative therapies are not effective, the proposal will not work
Over-prescribing of antibiotics	Homeopathy	As homeopathy is known to be ineffective, the proposal cannot work
Neglect of the whole person	Holistic therapies	If these therapies are not effective, the proposal is counterproductive
Not effective for chronic conditions	Integration of alternative therapies	If alternative therapies are not effective, the proposal will not work

Twenty-Seven
Ayurvedic Medicine

Ayurveda is an ancient Indian system of healthcare that emphasises prevention and health promotion. It considers the development of consciousness to be essential for optimal health and views meditation as the main technique for achieving this.[245] Treatments are highly individualised and include meditation, physical exercises, nutrition, forms of detox, relaxation, massage, medication, and yoga. Ayurvedic medicine strives for balance and claims that the suppression of natural urges leads to illness. The concepts of universal interconnectedness, the body's constitution (*prakriti*), and life forces (*doshas*) are important principles in Ayurvedic medicine. Treatments are often aimed at eliminating impurities from the body, increasing resistance to disease, reducing anxiety, and increasing harmony in life.

27.1. Charles involvement

In 2018, India's prime minister, Narendra Modi, paid a visit to the Science Museum in London where he inspected the '5,000 Years of Science and Innovation' exhibition. The event was hosted by Charles and included the announcement of new 'Ayurvedic Centres of Excellence', allegedly a 'first-of-its-kind' global network for evidence-based research on yoga and Ayurveda.[246] The first centre was said to open in 2018 in London. Funding was to come partly from the Indian government and partly from private donors. The central remit of the new initiative was reported to be researching the effects of Ayurvedic medicine.[247]

Dr Michael Dixon, the chair of the College of Medicine and Integrated Health (Chapter 23), commented:

This is going to be the first Ayurvedic centre of excellence in the UK. We will be providing, on the NHS, patients with yoga, with demonstrations and education on healthy eating, Ayurvedic diets, and massage including reflexology and Indian head massage. And all this will be subject to a research project led by Westminster University, to find out whether the English population will take to yoga and these sorts of treatments. Whether they will be helped by it and finally whether it will reduce the call on NHS resources leading to less GP consultations, hospital admissions and operations.[246]

On its website, the College of Medicine and Integrated Health announced that a memorandum of understanding with India's Ministry of Ayurveda, Yoga and Naturopathy, Unani, Siddha and Homoeopathy (AYUSH) had been signed 'to create centres of excellence in the UK... Dr Michael Dixon agreed the joint venture to provide the UK centres, which will offer and research traditional Indian medicine... The Indian government will match private UK donations to fund the AYUSH centres in the UK'.[248] In November 2019, the following press release by the President of India offered more details:

> The Prince of Wales called on the President of India, Shri Ram Nath Kovind, at Rashtrapati Bhavan today (November 13, 2019).
>
> Welcoming the Prince to India, the President congratulated him on his election as the head of the Commonwealth. He said that India considers the Commonwealth as an important grouping that voices the concerns of a large number of countries, including the Small Island Developing States.
>
> The President said that India and the United Kingdom are natural partners bound by historical ties and shared values of democracy, rule of law and respect for multi-cultural society. As the world's pre-eminent democracies, our two countries have much to contribute together to effectively address the many challenges faced by the world today.
>
> The Prince planted a Champa sapling — plant native to the subcontinent which has several uses in Ayurveda — in the

Herbal Garden of Rashtrapati Bhavan. He was taken around the garden and shown different plants that have medicinal properties. The Prince showed a keen interest in India's alternative model of healthcare.

The President thanked the Prince of Wales for his support for Ayurveda research. The Prince of Wales Charitable Foundation and the All India Institute of Ayurveda signed an MOU during the visit of Prime Minister Narendra Modi to the UK in April 2018. Under the MOU, the All India Institute of Ayurveda and the College of Medicine, UK will be conducting clinical research on Depression, Anxiety and Fibromyalgia. They will also be undertaking training programmes for the development of Standard Operating Protocol on 'AYURYOGA' for UK Health professionals.[249]

In 2019, Charles gave an address to a conference on yoga organised by 'The Yoga in Healthcare Alliance' and the 'College of Medicine and Integrated Health':

> The ancient practice of yoga has proven beneficial effects on both body and mind… For thousands of years, millions of people have experienced yoga's ability to improve their lives … The development of therapeutic, evidence-based yoga is, I believe, an excellent example of how yoga can contribute to health and healing. This not only benefits the individual, but also conserves precious and expensive health resources for others where and when they are most needed… I will watch the development of therapeutic yoga in the UK with great interest and very much look forward to hearing about the outcomes from your conference.[250]

In 2021, Charles told a yoga conference that doctors should work together with 'complementary healthcare specialists' to 'build a roadmap to hope and healing' after COVID-19. 'This pandemic has emphasised the importance of preparedness, resilience and the need for an approach which addresses the health and welfare of the whole person as part of society, and which does not merely focus on the symptoms alone. As part of that approach, therapeutic, evidenced-informed yoga can contribute to health and healing. By its very nature, yoga is an accessible practice which provides practitioners with ways to manage stress, build resilience and promote healing.'[251]

27.2. The evidence

There already exists a large body of research into Ayurvedic medicine; Medline, the largest databank for medical papers, lists in excess of 7,000 articles on Ayurveda and another 7,000 on yoga. Yet, encouraging evidence has remained scarce.

Ayurvedic remedies usually are mixtures of multiple ingredients and can consist of plants, animal products, and minerals. They often also contain toxic substances, such as heavy metals which are deliberately added in the ancient but erroneous belief that they can have positive health effects.[252] Other Ayurvedic remedies are known to cause liver damage.[253] In this context, it is interesting to note that the new centres of excellence are not destined to investigate the risks of Ayurvedic medicine, which arguably would be the most urgent topic of research.

Most of the existing studies of Ayurvedic treatments are methodologically weak. A Cochrane review, for instance, concluded that, '...there is insufficient evidence at present to recommend the use of these interventions in routine clinical practice...'[254] The efficacy of Ayurvedic remedies depends, of course, on the exact nature of the ingredients. Generalisations are therefore problematic. Promising findings exist only for a very small number of single ingredients, including Boswellia, Turmeric,[255] Frankincense,[256] and Andographis paniculata.[257]

Meditation is the term used for techniques that focus someone's mind on a particular object and are aimed at temporarily achieving a mentally clear and emotionally calm state. It has ancient roots and has been part of most religions, most notably Buddhism and Hinduism. As an alternative therapy, meditation is supposed to induce deep relaxation which, in turn, is said to have positive effects on a wide range of conditions. Even though there are many clinical studies and reviews of meditation, the evidence is far from strong, not least because of the methodological problems encountered and the frequently poor quality of these trials.

- One systematic review, for instance, concluded that 'there is some evidence that meditation is beneficial in improving quality of life in asthma patients. As two out of four studies in our review were of poor quality, further trials with better methodological quality are needed to support or refute this finding'.[258]
- Another systematic review found that 'at present there is not enough information available on the effects of meditation in haematologically-diseased patients to draw any conclusion'.[259]
- A further review concluded that 'Meditation programs... reduce multiple negative dimensions of psychological stress. Stronger study designs are needed to determine the effects of meditation programs in improving the positive dimensions of mental health as well as stress-related behavioral outcomes'.[260]
- Another review found that, 'as a result of the limited number of included studies, the small sample sizes and the high risk of bias, we are unable to draw any conclusions regarding the effectiveness of meditation therapy for ADHD. The adverse effects of meditation have not been reported. More trials are needed.'[261]
- And further review found that 'meditation interventions for older adults are feasible, and preliminary evidence suggests that meditation can offset age-related cognitive decline'.[262]

Only minor direct risks of meditation have been noted. One concern can be that, via meditation classes, consumers can be (and often are) recruited to some form of cult or sect.

Yoga has been defined in different ways in the various Indian philosophical and religious traditions. From the perspective of alternative medicine, it is a practice of gentle stretching exercises, breathing control, meditation, and lifestyles. The aim is to strengthen *prana*, the vital force as understood in traditional

Indian medicine. Thus, it is claimed to be helpful for most conditions affecting mankind.

There have been numerous clinical trials of various yoga techniques. They tend to suffer from poor study design and incomplete reporting and are thus not always reliable. Several systematic reviews have summarised the findings of these studies. An overview from 2010 included 21 systematic reviews relating to a wide range of conditions. Nine systematic reviews arrived at positive conclusions, but many were associated with a high risk of bias. Unanimously positive evidence emerged only for depression and cardiovascular risk reduction.[263] There is no evidence that yoga speeds the recovery after COVID-19 or any other severe infectious disease, as Charles seemed to suggest.

Yoga is generally considered to be safe. However, a large-scale survey found that approximately 30% of yoga class attendees had experienced some type of adverse event. Although the majority had mild symptoms, the survey results indicated that patients with chronic diseases were more likely to experience adverse events.[264] The warning by the Vatican's chief exorcist that yoga leads to 'demonic possession'[265] might not be taken seriously by yoga fans. Yet, some experts have warned that some yoga teachers try to recruit their clients into the more cult-like aspects of yoga.[266]

In essence, the evidence shows that some Ayurvedic treatments might have some potential but they are not without danger; and some of the risks are serious.

27.3. Consequences

When I last looked (January 2023), I could not find a trace of the 'Ayurvedic Centres of Excellence' that were announced with so much enthusiasm and fanfare three years ago (apart from an odd establishment that merely alludes to it).[267] This seems to suggest that the project has run into major difficulties. Or might it even need adding to the list of Charles' failures in the realm of alternative medicine?

Box 27

An example of an Ayurvedic therapy: marma massage

- Marma massage is said to be a deeply relaxing treatment.
- The marma therapist focuses on 107 assumed 'energy points' on the body.
- Marma massage is claimed to be 'one of the greatest healing secrets of Ayurveda'.
- Mosaraf Ali claims to be an expert in this therapy (Chapter 12).
- Charles initiated the only study of marma massage. It was conducted at Exeter and concluded that 'the effectiveness data showed no significant differences in changed scores'.[268]

Twenty-Eight

Social Prescribing

Social prescribing is a concept that emerged in the UK around the beginning of the century.[269] It aims to connect patients to different types of community support; for instance, social events, fitness classes, or social services. Trained professionals, called link workers, collaborate with healthcare providers such as GPs or community nurses offering referrals to these types of support. Social prescribing stems from the realisation that healthcare professionals cannot always address every need that their patients might have, but are nevertheless often in a key position to initiate activities to meet such needs.

A simple example is loneliness, which can cause stress, affect sleep, lifestyle, and physical or mental health. Doctors may not be able to cure this problem; however, link workers might provide support. They can learn about patients' requirements, might motivate patients, and refer them to helpful resources in the community (Box 28).[270]

28.1. Charles' views

In 2021, Charles published an article in the *Future Healthcare Journal* in which he elaborated on social prescribing (or social prescription, as he calls it).[271] Below is the key passage from this paper (the numbers in square brackets refer to my comments below):

>...For a long time, I have been an advocate of what is now called social prescription and this may just be the key to integrating the biomedical, the psychosocial and the

environmental, as well as the nature of the communities within which we live and which have such an enormous impact on our health and wellbeing.[1] In particular, I believe that social prescription can bring together the aims of the health service, local authorities, and the voluntary and volunteer sector. Biomedicine has been spectacularly successful in treating and often curing disease that was previously incurable. Yet it cannot hold all the answers, as witnessed, for instance, by the increasing incidence of long-term disease, antibiotic resistance and opiate dependence.[2] Social prescription enables medicine to go beyond pills and procedures and to recognise the enormous health impact of the lives we lead and the physical and social environment within which we live.[3] This is precisely why I have spent so many years trying to demonstrate the vitally important psychosocial, environmental and financial added value of genuinely, sustainable urban planning, design and construction.[4].

There is research from University College London, for instance, which shows that you are almost three times more likely to overcome depression if you have a hobby.[5] Social prescription enables doctors to provide their patients with a bespoke prescription that might help them at a time of need...

When we hear that a quarter of 14–16-year-old girls are self-harming and almost a third of our children are overweight or obese, it should make us realise that we will have to be a bit more radical in addressing these problems.[5] And though social prescription cannot do everything, I believe that, used imaginatively, it can begin to tackle these deep-rooted issues.[6] As medicine starts to grapple with these wider determinants of health,[7] I also believe that medicine will need to combine bioscience with personal beliefs, hopes, aspirations and choices.[8]

Many patients choose to see complementary practitioners for interventions such as manipulation, acupuncture and massage.[9] Surely in an era of personalised medicine, we need to be open-minded about the choices that patients make and embrace them where they clearly improve their ability to care for themselves?[10] Current NHS guidelines on pain that acknowledge the role of acupuncture and mindfulness may lead, I hope, to a more fruitful discussion on the role of complementary medicine in a modern health service.[11] I have always advocated 'the best of both worlds',[12] bringing evidence-informed[13] conventional and complementary medicine together and avoiding that gulf between them, which leads, I understand, to a substantial proportion of patients feeling that they cannot discuss complementary medicine with their doctors.[14]

I believe it is more important than ever that we should aim for this middle ground.[15] Only then can we escape divisions and intolerance on both sides of the conventional/complementary equation where, on the one hand, the appropriate regulation of the proven therapies of acupuncture and medical herbalism[15] is opposed while, on the other, we find people actually opposing life-saving vaccinations. Who would have thought, for instance, that in the 21st century that there would be a significant lobby opposing vaccination, given its track record in eradicating so many terrible diseases and its current potential to protect and liberate some of the most vulnerable in our society from coronavirus?[16]

28.2. My comments

My comments are as follows:

1. Charles might be a little generous to his own vision here; the idea of social prescribing is not the same as the concept of integrated medicine (Chapter 13).
2. To date, there is no good evidence that social prescribing will reduce 'long-term disease, antibiotic resistance and opiate dependence'.
3. Here Charles produces a classic 'strawman fallacy' (Chapter 30). Medicine is, of course, more than pills and procedures.
4. Charles has not so much 'demonstrated' the importance of 'psychosocial, environmental and financial added value of

genuinely, sustainable urban planning, design and construction' as talked about it.
5. That does not necessarily mean that social prescribing is effective; correlation is not causation!
6. Currently, there is no reliable evidence that social prescribing is effective against self-harm or obesity.
7. Medicine has been trying to grapple with 'wider issues' for centuries. The concept of social prescribing was developed by medical experts and is merely another example of this fact.
8. Medicine has done that for many years but it always had to be mindful of the evidence base. Adopting interventions without good evidence that they do more good than harm would render healthcare less safe and less effective.
9. Many patients also choose to smoke, drink alcohol, sky-dive, etc. Patient choice is not a reliable indicator for efficacy or harmlessness.
10. Yes, we should embrace treatments that clearly improve patients' ability to care for themselves. However, the evidence all too often fails to show that alternative therapies improve anything.[29]
11. As we have seen throughout this book, this discussion has been going on for decades and was not always helped by Charles (Chapter 6).
12. The best of both worlds entails those treatments that demonstrably do more good than harm; they are not necessarily therapies that Charles likes best (Chapter 13).
13. 'Evidence-informed' is an interesting term that already cropped up in the previous chapter and is fast becoming a popular term with advocates of alternative medicine. Modern medicine strives to be evidence-based. Evidence-informed seems to imply that evidence is less important than adhering to evidence-based medicine. Why? Perhaps because, for alternative medicine, the evidence is largely not convincingly positive?
14. If we want to bridge the gulf, we foremost require sound evidence. Today, there is plenty of such evidence.[29] The

problem is that it mostly does not show what Charles assumes or wants.
15. As I already noted in Chapter 8 of this book, even the best regulation of nonsense must result in nonsense.
16. The anti-vaccination sentiments originate to a worryingly large extent from the realm of alternative medicine.[272]

These points might seem perhaps a little petty; much more important, to my mind, is the fact that Charles uses the simple 'bait and switch' trick to sell his message. Bait and switch is 'a morally suspect sales tactic that lures customers in with specific claims about the quality or low prices on items that turn out to be unavailable in order to upsell them on a similar, pricier item. It is considered a form of retail sales fraud, though it takes place in other contexts'.[273] The bait, in this case, is 'social prescribing' and the switch is alternative medicine. Social prescribing might turn out to be a helpful concept (at present, the evidence is far from compelling),[274] but it is used here for no other purpose than to smuggle mostly unproven alternative therapies into the mainstream of routine healthcare. Just like the concept of integrated medicine (Chapter 13), social prescribing serves Charles as a smokescreen behind which he wants to incorporate treatments into medicine which otherwise would not pass muster. The disservice to public health of using social prescribing in this way is obvious: healthcare would not become better but worse. To put it in a nutshell: social prescribing of nonsense must result in nonsense.

28.3. Consequences

Social prescribing is a relatively new and currently fashionable idea. The first articles in medical journals emerged only about 15 years ago. The concept seems to make sense and its aims are laudable. Whether it will turn out to benefit public health largely depends on how it is implemented. At present, there is a lack of reliable evidence and we simply cannot tell. Dr Dixon (Box 10), who has recently become an enthusiastic promoter of social prescribing, admits that 'there may be gaps in the current evidence base (as with any fast-moving social movement) but these will surely be more than compensated with faith, courage and a renewal of medical ideals tempered with a little common sense'.[275] NHS England concurs: 'there is a need for more robust and systematic evidence on the effectiveness of social prescribing.'[276]

Social prescribing faces numerous challenges, and faith, courage, and common sense might not suffice to overcome them.[277] If our governments decide to use social prescribing for obfuscating the fact that our health services are under-staffed and under-supported, a focus on social prescribing might turn out to be detrimental. If doctors use it to delegate core medical tasks to link workers, they might erode their own role in healthcare. If social prescribing is used to integrate unproven treatments into routine care, it is unlikely to be of benefit to patients.

Box 28

Examples of social prescribing

- People who feel at a loose end might be referred to book clubs, music groups, theatre guilds, etc.
- Patients suffering from arthritis can be referred to an arthritis support hub.
- People who are lonely can be referred to local groups where they can meet new people and make friends.
- Patients who are obese can be referred to keep fit initiatives and might be taught about healthy eating and healthy cooking.
- People who experience stress or anxiety can be referred to relaxation classes.

Twenty-Nine

Homeopathy

Amongst all alternative therapies, homeopathy stands out as Charles' favourite, and has for many years been at the centre of many of his promotional efforts. I have therefore saved it to last.

Homeopathy was invented some 200 years ago by the German physician, Samuel Hahnemann (1755–1843). It relies on two main principles:

- Like cures like, i.e. coffee keeps us awake, therefore homeopathic coffee should cure sleeplessness.
- Diluting remedies to the point where no active molecule is left renders them not weaker but stronger.

Both of these principles fly in the face of common sense and scientific fact. At the time when Hahnemann invented homeopathy, our understanding of the laws of nature was woefully incomplete, and therefore Hahnemann's ideas seemed less implausible then than they do today. Many of the conventional treatments of 200 years ago were more dangerous than the disease they were supposed to cure. Consequently, homeopathy proved to be better than 'allopathy' (a term coined by Hahnemann to denigrate conventional medicine), and Hahnemann's treatments were an almost instant worldwide success. Today, there is a broad consensus that homeopathy is a placebo therapy. NHS England, for instance, states: 'There's no good-quality evidence that homeopathy is effective as a treatment for any health condition.'[278]

29.1. Charles' views

Whenever Charles promotes alternative medicine, he is likely to mention or think of homeopathy. His enthusiasm for homeopathy is perhaps less surprising if we consider the historical context. Generations of royals have favoured homeopathy.

The royal protection of homeopathy can be seen in numerous guises:

- Not least because of royal influence did homeopathy become part of the NHS from its beginning in 1948.
- After managing to get osteopathy and chiropractic regulated by statute, Charles had planned to do the same with homeopathy. Jonathan Dimbleby wrote in 1995: 'It is now hoped that a Homeopathy Bill will be laid before the House in 1995 or 1996.'[6]
- Charles advocates homeopathy not just for humans but also for animals. Farmers in the UK, for instance, are being taught how to treat their livestock with homeopathy 'by kind permission of His Royal Highness, The Prince of Wales'.[279]
- Ainsworths, the UK homeopathic pharmacy, carries Charles' royal warrant.[280]
- Charles often supports homeopathic events with his presence and speeches. For instance, he opened the Glasgow Homeopathic Hospital.[281]
- The Smallwood Report commissioned by Charles concluded that millions of pounds could be saved, if only the NHS used more homeopathy (Chapter 15).
- The College of Medicine and Integrated Medicine (Chapter 23) of which Charles now is a patron regularly promotes homeopathy.[282]
- Charles regularly lobbied politicians urging them to make more homeopathy available via the NHS (Chapter 22).
- In 2019, the Faculty of Homeopathy announced that His Royal Highness the Prince of Wales (as he was at the time) had accepted to become Patron of the Faculty of Homeopathy.[283]

In his book *Harmony*,[2] Charles elaborates:

> ...[O]ne of the big arguments used against homeopathy is that it does not really work medically. The criticism is that people simply believe they are going to feel better and so they *think* they are better. They have responded to the so-called 'placebo effect'. It is for this reason that critics argue that it is a trick of the mind and its remedies are nothing more than sugar pills. What none of those who take this view ever seem to acknowledge is that these remedies also work in animals, which are surely unlikely to be influenced by the placebo effect. I certainly remember that when I started to introduce homeopathic remedies on the Duchy Home Farm, farm staff who had no view either way reported that the health of an animal that had been treated had improved so I wonder what it is that prevents the medical profession from even considering the evidence that now exists of trials of homeopathic treatments carried out on animals? It is not the quackery they claim it to be. Or if it is, then I have some very clever cows in my shed!

As we will see in a moment, 'the evidence that now exists of trials of homeopathic treatments carried out on animals' is precisely what doctors, vets, scientists, and other experts considered and what made them reject homeopathy as a mere placebo therapy.

29.2. The evidence

Homeopathic remedies can be based on plants, or on any other material (a famous remedy is made of the Berlin Wall),[284] or no material at all (Box 29). They are typically so dilute that they contain not a single molecule of the substance advertised on the bottle. The most frequently used dilution (homeopaths call them 'potencies') is called a 'C30'; a C30-potency has been diluted 30 times at a ratio of 1:100. This means that one drop of the starting material is dissolved in 1,000,000,000,000,000,000,000,000,000,000, 000,000,000,000,000,000,000,000,000 drops of diluent (usually a water/alcohol mixture) — and that equates to less than one molecule of the original substance per all the molecules of the universe. Homeopaths know that, of course, and claim that their remedies work via some undefined 'energy' or 'vital force' and that the process of preparing the homeopathic dilutions (which involves vigorously shaking the mixtures at each dilution step) transfers this 'energy' or information from one to the next dilution. They also believe that the process of diluting and agitating their remedies, which they call potentisation, renders them not less but more potent.[285]

Homeopathic remedies are usually prescribed according to the 'like cures like' principle: if, for instance, a patient suffers from watery eyes, a homeopath might prescribe a remedy made of onion, because onion make our eyes water. Or if the Berlin Wall caused relationship problems, the homeopathic version will cure them. This and all other assumptions of homeopathy contradict the known laws of nature. In other words, we do not fail to comprehend how homeopathy works, as many enthusiasts claim, but we understand that homeopathy cannot work unless the known laws of nature are wrong.

Today, around 500 clinical trials of homeopathy as a treatment for various diseases and symptoms have been published (the first was published 1835 in Nuremberg by a group of 'truth loving men'; its findings were squarely negative).[286] With this number of studies, it is to be expected that some come out positive purely by chance. These are the trials that homeopaths often use to persuade the public that homeopathy works after all. However, the totality of the reliable evidence fails to show that highly diluted homeopathic remedies are more than placebos.[287] Numerous official statements from various countries confirm the absurdity of homeopathy, for instance:

- 'The principles of homeopathy contradict known chemical, physical and biological laws and persuasive scientific trials proving its effectiveness are not available' (Russian Academy of Sciences, Russia).
- 'Homeopathy should not be used to treat health conditions that are chronic, serious, or could become serious. People who choose homeopathy may put their health at risk if they reject or delay treatments for which there is

good evidence for safety and effectiveness' (National Health and Medical Research Council, Australia).
- 'Homeopathic remedies don't meet the criteria of evidence-based medicine' (Hungarian Academy of Sciences, Hungary).
- 'The incorporation of anthroposophical and homeopathic products in the Swedish directive on medicinal products would run counter to several of the fundamental principles regarding medicinal products and evidence-based medicine' (Swedish Academy of Sciences, Sweden).

Yet, Charles is right: many patients (human or animal) undeniably do get better after taking homeopathic remedies. The best evidence available today clearly shows that this improvement is unrelated to the homeopathic remedy *per se*. In humans, it is the result of a lengthy, empathetic, compassionate encounter with a homeopath, a placebo-response, or other factors which experts often call 'context effects'.[288]

But what about homeopathy for animals? Charles claimed in his book and on many other occasions that there is good evidence for it and stated that he has been successfully using homeopathy on his cows (Chapter 25). Veterinary homeopathy is indeed a well-known phenomenon. Ever since Samuel Hahnemann gave a lecture on the subject in the mid-1810s, homeopathy has been used for treating animals. Initially, veterinary medical schools tended to reject homoeopathy as implausible, and the number of veterinary homeopaths remained small. In the 1920s, however, veterinary homoeopathy was revived in Germany, and in 1936 members of the 'Studiengemeinschaft für Tierärztliche Homöopathie' (Study Group for Veterinary Homoeopathy) started to investigate homeopathy systematically. Today, it has again become popular.

Yet, popularity is not the same as efficacy. A 2015 systematic review (published by ardent homeopaths) tested the hypothesis that the outcome of veterinary homeopathic treatments is distinguishable from placebos. A total of 15 trials could be included, but only two comprised reliable evidence without overt vested interest. The authors concluded that there is 'very limited evidence that clinical intervention in animals using homeopathic medicines is distinguishable from corresponding intervention using placebos'.[289] A more recent systematic review compared the efficacy of homeopathy to that of antibiotics in cattle, pigs, and poultry. A total number of 52 trials were included of which 28 were in favour of homeopathy and 22 showed no effect. No study had been independently replicated. The authors concluded that 'the use of homeopathy cannot claim to have sufficient prognostic validity where efficacy is concerned'.[290] It is thus clear that the findings from veterinary studies mirror those of the human investigations: highly diluted homeopathic remedies are pure placebos, and the positive effects that are being observed are due to phenomena that are unrelated to homeopathy *per se*.

But at least they are safe! Charles, like many other homeopathy enthusiasts, assumes that homeopathy is entirely harmless. This is sadly not true either. Whenever homeopaths advise their patients (as they often do) to forgo effective conventional treatments, they are likely to do significant harm. This phenomenon is best documented in relation to the advice of many homeopaths against immunisations.[272] Other homeopaths and manufacturers of homeopathics go even further and advocate homeopathy as a treatment of cancer.[291,292]

29.3. Consequences

Today even mainstream UK media like *The Times* seem to have run out of patience with Charles' relentless promotion of homeopathy:

> Prince Charles' endorsement [of homeopathy] might turn out to be a poisoned chalice. His reputation in science is not exactly the best, and his patronage will simply re-emphasise the many negative verdicts of independent experts on homeopathy. The European Academies Science Advisory Council, for instance, stated recently this: '…we acknowledge that a placebo effect may appear in individual patients but we agree with previous extensive evaluations concluding that there are no known diseases for which there is robust,

reproducible evidence that homeopathy is effective beyond the placebo effect'.[293]

Charles' defence of the indefensible has cost him credibility not merely in the scientific community, but even amongst many advocates of alternative medicine. Charles' unwavering advocacy of homeopathy in the face of the mounting evidence against it also highlights his disregard of science and his determination to stand in the way of progress.

Not only that, but Charles' passion for homeopathy is also the best example for showcasing the failure of his promotional activities. When Charles first sided with homeopathy, the UK had five homeopathic NHS hospitals; today, there are none that carry the name. When he started on his mission to integrate more homeopathy into routine healthcare, the NHS reimbursed the costs of homeopathy; today, this is no longer the case. Some people claim that this decline is due to the work of a highly organised and well-funded group of sceptics. But this is quite simply not true: firstly, the UK sceptics are neither highly organised nor well-funded (contrary to Charles' various initiatives), and secondly, the cuts are predominantly due to the fact that there is no good evidence for homeopathy and that, in the age of evidence-based medicine, it is no longer felt ethical to carry on funding disproven treatments.

Charles acts on his intuition, his beliefs, his convictions. He seems entirely immune to evidence that does not confirm his creed. In that, he can become a true enemy of the age of reason (Chapter 24). He has no competence in science or medicine and takes advice only from people who are of his opinion in the first place. But he can do what he wants, people might say. I agree without hesitation. He can surely have whatever opinion he wants to have, but, just as surely, he cannot have his own facts. Modern healthcare is not based on creed; in the best interests of the patient, it must be based on evidence. This evidence evolves, is scrutinised and discussed until a consensus is reached. When Charles contradicts the consensus, when he uses his influence to interfere with our healthcare, and when he pretends his opinion amounts to evidence, he stands in the way of progress.

Negative examples by prominent people inevitably endanger others. A case in point is Charles' recent coronavirus illness. When he recovered quickly and apparently unscathed, there were immediately those, including the Indian Minister of State for Ayush, Shripad Naik, who claimed this was due to homeopathy.[294] Clarence House quickly denied this, but the notion made headlines in India and the officials decided to use homeopathy to keep the pandemic at bay. Unquestionably, this decision cost uncounted lives during the wave of COVID-19 infections that followed this disastrous decision.

Box 29

Examples of homeopathic remedies that are not based on any material at all

- Blue (colour)
- Eclipse Totality
- Electricitas (Electricity — 80,000 volts)
- Electricitas (High Voltage Pylon)
- Green (colour)
- Halogen light
- Indigo (colour)
- Laser — red (Diode Laser Red)
- Laser Beam (2940 nm)
- LED (white) (White L(ight) E(mitting) D(iode) Light)
- Light (energy saving bulb)
- Luna (Moonlight)
- Microwave 750 MHz
- Milky Way (Essence)
- Mobile phone (Eising)
- Mobile Phone 1800MHz
- Mobile Phone 900Mhz
- Mobile Phone Mast G3
- Polaris (North star)
- Purple (Colour)
- Radiation Combination (Guild Radiation Combination)
- Rainbow (Spectrum)
- Red (colour)
- Red (A. Wauters)
- Sol Africana
- Sol Australis (Sunlight — Australia)
- Sol Britannic (Sunlight — British)
- Stonehenge (Emanation)
- Sunlight Blue (Prismatic blue from sunlight)
- Sunlight Green (Prismatic green from sunlight)

- Sunlight Orange (Prismatic orange from sunlight)
- Sunlight Purple (Prismatic purple from sunlight)
- Sunlight Red (Prismatic red from sunlight)
- Sunlight Yellow (Prismatic yellow from sunlight)
- Ultrasound (General)
- Ultrasound (Vaginal)
- Ultraviolet Light
- Vacuum
- Wind (South-West)
- X-Ray

Thirty
Final Thoughts

In this last chapter, I will briefly summarise the essence of the previous ones and offer some final thoughts about Charles' passion for alternative medicine.

30.1. Extrapolations

At boarding school, Charles was reading the books of van der Post and started dreaming of different horizons. He was drawn toward spiritualism and, while at Cambridge, dabbled in parapsychology. His fellow students at the time were revolting against the establishment, and he must have felt like joining in. But how could he? He was the heir to the throne! Who could be a more perfect representative of the establishment than Charles?

Then Charles met the charismatic Laurens van der Post in person and soon fell under the spell of the teller of tall tales. Charles was looking for the meaning of life and Laurens was skilled at offering him a 'missing dimension'. Laurens' influence emboldened the deeply insecure Charles to follow his special interests, one of which would become a life-long passion: alternative medicine.

But Charles' degree in arts left him ill-equipped to comprehend science or medicine, so Laurens convinced him that his royal intuitions came 'from a far deeper source than conscious thought'. And soon Charles found himself on a mission: he revolted against the establishment—albeit the medical one—in his very own and privileged way. In 1982, he lectured the medical elite that 'today's unorthodoxy is probably going to be tomorrow's convention' and later he rejoiced that 'I was absolutely astonished to find what a reaction it had caused amongst the medical establishment'. Even in 2005, Charles' anti-establishment sentiments were still tangible when he wrote to then Health Secretary, Alan Johnson, that 'despite waves of invective over the years from parts of the medical and scientific establishment', he continued to lobby 'because [he] cannot bear people suffering unnecessarily when a complementary approach could make a real difference'.

At an early stage of Charles' life, the dice had been cast and his destiny as a reformer of medicine had become evident. Charles felt he could 'restore the human being to a lost natural aspect' of which health was to play a key role, and he began to 'regard his own instincts as revealed truths superior to those of professionals and experts'.[295]

30.2. The facts

I am, of course, extrapolating; we have no means of knowing whether things did happen quite like that. But, whatever motivated Charles, the facts are plain to see: Charles became the undisputed champion of alternative medicine, much admired by alternative medicine fans and castigated by scientists and sceptics. In this role, he seemed to indiscriminately value any odd alternative, as long as it was not part of the medical establishment.

In conventional healthcare the main criteria for considering the value of a therapy include:

- Plausibility
- Effectiveness
- Safety

None of these seem to matter to Charles.

- He is a staunch defender of homeopathy, arguably **the least plausible** of all alternative therapies.
- He advocates iridology, arguably **the least effective** diagnostic method imaginable.
- He defends chiropractic, arguably **the least safe** treatment in alternative medicine.

Charles' passion for alternative medicine knows no bounds, it ignores conventions because it is not based on reason but on intuition.

30.3. Charles' choice of alternative medicine

Charles seems to support alternative modalities mainly for one reason: because they are alternative! It almost appears that any method which might upset the medical establishment will be approved by Charles. As we have seen in the preceding chapters, in the past, Charles has actively promoted:

- Acupuncture
- Aromatherapy
- Ayurveda
- Chiropractic
- Detox
- Gerson therapy
- Herbal medicine
- Homeopathy
- Iridology
- Marma massage
- Massage therapy
- Osteopathy
- Pulse diagnosis
- Reflexology
- Tongue diagnosis
- Traditional Chinese Medicine
- Yoga

The types of alternative medicine supported by Charles are strikingly heterogeneous. Homeopathy has nothing in common with herbalism; osteopathy has nothing in common with acupuncture; chiropractic has nothing in common with aromatherapy, etc., etc. Crucially, the assumptions of one alternative therapy are not compatible with those of another. They cannot all be correct; more likely they are all incorrect. Charles' choice of therapeutic and diagnostic methods seems confusing indeed. At first glance, the only common denominator seems to be the fact that they are not part of conventional medicine. Yet, at second glance, two communalities do emerge.

30.4. Charles' selection criteria

Intriguingly, there is a total absence of any 'high tech', modern, or synthetic version of alternative medicine in the list above. Yet many such alternative modalities do exist. For example, bioresonance, chelation therapy, hair analysis, neural therapy, live blood analysis, orthomolecular medicine, and Vega testing are all well-known alternative modalities which are missing from Charles' list — and they all involve modern technologies.[29]

It seems, therefore, that Charles' focus is on alternative methods which are firstly old and secondly natural. In an editorial, Charles once explained that integrated medicine is 'the kind of care that integrates the best of new technology and current knowledge with ancient wisdom' (Chapter 13). His book *Harmony* is claimed to be 'practical guide to what we have lost in the modern world, why we have lost it and how easily it is to rediscover...'[2] In it, Charles states that 'even as a teenager I felt deeply disturbed by what seemed to have become a dangerously short-sighted approach. I could not help feeling that in whichever field these changes were taking hold, with industrialized techniques replacing traditional practices, something very precious was being lost.' In the same book, Charles writes that 'if we ignore Nature, everything starts to unravel'. And in his WHO lecture (Chapter 17), his preference for the natural and ancient traditions as well as his disdain for modern options are expressed openly:

> Many of today's complementary therapies are rooted in **ancient traditions** that intuitively understood the need to maintain balance and harmony with our minds, bodies and the **natural** world. Much of this knowledge, often based on oral **traditions**, is sadly being lost yet orthodox medicine has so much to learn from it. It is tragic, it seems to me — and indeed to many people who have studied this whole area — that in the **ceaseless rush to 'modernize'**, many beneficial

approaches, which have been tried and tested and have shown themselves to be effective, have been cast aside because they are deemed to be 'old-fashioned' or 'irrelevant' to today's needs.

30.5. Fallacies

A striking feature of Charles' choices and pronouncements is his abundant use of logical fallacies, i.e. unsubstantiated assertions delivered with such conviction that they almost sound like proven facts.

The appeal to tradition is a common logical fallacy based on the assumption that a traditional practice is better than a newer option. If a therapy has survived for many decades, it must be fine, according to this fallacious argument—otherwise it would have been discarded long ago. And many alternative treatments are indeed quite old:

- Acupuncture has a history of 2,000 years.
- Herbal medicine is even older.
- Homeopathy has survived more than 200 years.
- Chiropractic and osteopathy are more than 120 years old.
- Etc., etc.

But the notion that a long history of usage is a substitute for scientific evidence is not just fallacious, it can also be dangerous. There are many historical examples to demonstrate this. For instance, blood letting was used in many cultures for hundreds of years. Today we know that it not only failed to cure the sick but also hastened the death of millions.

The fact that a therapy or diagnostic method has been around for centuries might merely signify that it emerged when our understanding about the human body was woefully incomplete. Seen from this perspective, a long history of usage should prompt us to question the plausibility of the assumptions underlying that modality.

The appeal to nature is the fallacious belief that something must be effective and safe because it comes from 'Mother Nature'. Few qualities attract people more than that of being natural. We prefer natural food to processed food; we prefer natural fibre and clothes to synthetic materials; we prefer natural behaviour to pretence; we prefer a natural environment to an artificial one; we prefer natural cosmetics to synthetic products; and, of course, we prefer natural medicines to synthetic ones. Chemicals are bad for us, according to this fallacious argument. Alternative medicine is therefore invariably promoted as being natural. But, if we think more critically about the claim, we find that there is little truly natural about alternative therapies, for example:

- Acupuncture involves the unnatural process of sticking needles into the skin.
- Herbal supplements are highly processed.
- Homeopathy often employs artificial materials like, for instance, the Berlin Wall.[284]
- Spinal manipulation moves joints outside their natural range of motion.

The assumption that natural necessarily means safe is equally fallacious. A virus, a tsunami, a flash of lightning, and many other natural phenomena are anything but harmless. Even alternative therapies labelled as natural are not free of risks, for instance:

- Acupuncture can cause death through cardiac tamponade.[296]
- Some herbal remedies can cause liver failure.[297]
- Chiropractic spinal manipulations can cause a stroke followed by death.[52]

In alternative medicine, the term 'natural' is little more than an advertising slogan without substance. Alternative therapies are not truly natural, and natural does not necessarily mean harmless. In fact, most forms of alternative medicine that are deemed to be natural are neither effective nor safe.

The strawman fallacy is an argument in which an opponent's position is misrepresented in order to be more easily refuted. In discussions about alternative medicine, the strawman fallacy is used regularly for misrepresenting conventional medicine; the aim is to make it look unattractive. Charles employs this fallacy almost every time he talks about the insufficiencies of modern medicine which, he believes, need to be corrected by integrating alternative medicine. For instance, in his recent journal article he states that 'social prescription enables medicine to go beyond pills and procedures and to recognise the enormous health impact of the lives we lead and the physical and social environment within which we live'.[271] Another example is his 2021 statement related to yoga (Chapter 27): 'This pandemic has emphasised the importance of preparedness, resilience and the need for an approach which addresses the health and welfare of the whole person as part of society, and which does not merely focus on the symptoms alone.' He effectively misinterprets conventional medicine by implying that it is not about the welfare of the whole person but merely about treating symptoms.

I suspect that Charles knows very well that medicine has always been more than pills and procedures. It includes care, compassion, dialogue, empathy, etc. Yet Charles chooses to misrepresent it as being devoid of such elements so that his notion of integrated medicine appears more attractive (Chapter 13). He counts on the possibility that those who read his text do not see through his fallacious argument—and sadly, many don't.

The 'post hoc ergo propter hoc' fallacy refers to the assumption that two events that follow each other in time are causally related. We all know that the crowing of the cock at dawn is not the cause of the sun rising shortly after. Yet, many of us assume that the treatment we used yesterday is the cause of the improvement we experience today. Charles was quoted stating that 'there are different types of evidence, and the evidence of experience is just as important as scientific evidence'.[20] Similar notions emerge often from his writings: in his book *Harmony*,[2] for example, he elaborates on the reaction of one of his farm animals to a homeopathic treatment and concludes that the treatment was the cause of the improvement (Chapter 25). This is a classic 'post hoc' fallacy: the experience of an improvement can have several causes that are unrelated to the treatment (for instance, the cow might have gotten better because of the natural history of the disease). Since this is undeniably so, the evidence of experience is not relevant for establishing evidence about the effectiveness of medical interventions (Chapter 24).

The non sequitur fallacy is an argument that uses statements which do not follow the fundamental principles of logic and therefore arrive at conclusions that do not follow logically from what preceded. Charles seems prone to this type of fallacy.

In a way, his whole idea of integrated medicine is a 'non sequitur'. Charles argues that, because conventional medicine is imperfect (which is true), we should add unproven alternative treatments to it. He does not consider that this move would render conventional medicine all the more imperfect (Chapter 13). When Charles talks about antibiotic resistance, he makes his usage of the non sequitur fallacy even more obvious (Chapter 26). He correctly points out that we have overused antibiotics to an extent that it is now endangering our health. He then suggests that homeopathy is the solution to the problem. While his first argument is correct, the second does not follow logically from the first.

30.6. The tangible results of Charles' promotion of alternative medicine

Charles has been promoting alternative medicine for almost half a century. Considering his influence, one would expect that his activities have resulted in a respectable range of successes and tangible changes to public health in the UK and perhaps even beyond. Yet, if we critically evaluate the pertinent developments, this expectation is not realised.

- When Charles started campaigning for homeopathy, there were five NHS homeopathic hospitals in the UK; today there are none.
- Charles' attempts to get UK homeopaths, herbalists, and acupuncturists regulated by statute failed.
- The statutory regulation of chiropractic and osteopathy did not result in more acceptance of manual therapies within the NHS, nor in better evidence in support of these treatments (Chapter 8 and 9).

- Charles' model hospital of integrated medicine never materialised (Chapter 12).
- His Foundation of Integrated Health closed amid scandal (Chapter 13).
- Despite the Smallwood Report (Chapter 16), the NHS's use of alternative medicine has declined.
- NICE guidelines rarely recommend alternative therapies.[298]
- The findings of the 'Get Well Study' were not implemented in any region of the UK (Chapter 19).
- The Duchy Detox Tincture was swiftly taken off the market (Chapter 21).
- Charles' promotion of alternative medicine via his 'spider memos' was largely ignored by health politicians.[299]
- Reimbursement of homeopathy by the UK public purse ceased in 2018.[300]
- Charles' public image as a champion of progress in healthcare is in tatters.[301]

30.7 Charles' alternative

It is clear to many observers that Charles has the urge to make a positive contribution to the future of his country. Most agree that he is full of good will. In some areas, for example the Prince's Trust,[302] he was highly successful in his endeavour. In the field of alternative medicine, however, success has evaded him. One might ask, therefore, how he could have channelled his enthusiasm, influence, and hard work in a more productive direction. In my view, this would not have been difficult and could have been achieved by operating along the following lines:

- Work not against but alongside the medical and scientific establishment.
- Involve some of the country's top scientists.
- Raise sufficient funds for rigorous research projects conducted at leading universities.

- Encourage his team of science advisors to defend unpopular views and, if necessary, contradict Charles' views.
- Focus on treatments that are biologically plausible and supported by encouraging evidence, e.g. rational phytotherapy (Chapter 15).
- Make sure that the potential of harm of alternative medicine is fully investigated and the findings are adequately publicised.
- Become a defender of science and reason.

Some these principles are not all that dissimilar to those of the US Bravewell Collaborative (Chapter 20). Charles would only have needed to follow their example. It seems that he and his advisors did not consider this to be viable.

As he became king, Charles could have looked back at his activities around alternative medicine in the knowledge that—as with some of his other 'good causes'—he has provided tangible benefits for the people. Many of the negative headlines that Charles had to endure about his involvement in alternative medicine could have been different (Box 30), his reputation within the world of science would be intact, and the alternative medicine community might respect him even more.

30.8. The future

According to his own statement, Charles will stop his lobbying once he is king: when asked if his campaigning would carry on when he is king, Charles replied, 'No, it won't. I'm not that stupid.'[303] If that happens, alternative medicine will have lost one of its most enthusiastic supporters. In this case, I will look back on this period with a degree of sadness.

Despite everything, I still believe that alternative medicine has a few hidden gems to discover. To find them, we foremost need good science. To conduct the research, we need people with influence to support it. Charles could have so easily been that person. Instead, he took consistently poor advice and chose to follow a different path. He pursued a largely anti-science agenda and promoted the uncritical integration of unproven treatments into the NHS. In this way, I am afraid, he became an obstacle to progress in healthcare and generated more harm than good. My predominant feeling about that is sadness over a missed opportunity.

Box 30

Four examples of actual and fictitious headlines about Charles' passion for alternative medicine

Headline as published	Headline as it could/should have been
Prince Charles rejected by experts before Gwyneth Paltrow's long Covid row: 'Witchcraft'[304]	Prince Charles recognised as a supporter of sound science in alternative medicine
Prince Charles the 'snake-oil salesman'[305]	Prince Charles, the defender of reason
Prince Charles' therapist sued by amputee[306]	Nobel prize winner joins Charles' inner circle of advisors
Prince Charles lobbied the Prime Minister in support of 'alternative medicines', 'black spider' letters show[307]	Charles uses his influence to create more centres of excellence to study alternative medicine

Thirty-One

Long Live the King!

As mentioned in Chapter 5, Laurens van der Post expressed his hope that Charles would never become King. Van der Post feared that this would imprison him, stating that Charles 'is a natural Renaissance man'. Yet, now Charles has become King we may well ask what consequences his ascension to the throne will have for alternative medicine, for the royal family, and for the monarchy.

As King, Charles is supposed to be neutral and stay out of politics. Does that mean he will totally abandon his previous interests? Palace sources have already indicated that the King will continue to champion the environment while on the throne.[308] This surely must be applauded. But what about alternative medicine? More promotion of medical nonsense would certainly not be in the interest of public health.

Shortly after the Queen's passing, a spokesperson for Buckingham Palace provided the following statement: 'As Prince of Wales, His Royal Highness came to the view that complementary medicine could play a role in healthcare, as long it was integrated with conventional treatments, a position he reached after years of talking to experts in many different areas of medicine.'[309] This seems to indicate that Charles intends to carry on defending homeopathy and similarly dubious treatments.

The fact that the scientific case for the efficacy and safety of these therapies has recently become less and less convincing or even outright negative (Table 1) is unlikely to deter Charles. His support might now become less visible and more discrete but, at the same time, it will be much more influential. 'The monarchy has a huge amount of indirect power in that it can influence public opinion on a matter, which is arguably more important than lobbying ministers', stated Kate Williams, a leading royal historian.[310]

There are early signs of Charles' continuation of support of alternative medicine. For example, since ascending the throne, he remained patron of several key organisations in this area, e.g. the Faculty of Homeopathy (Chapter 29), the College of Medicine and Integrated Health (Chapter 23), and the homeopathic pharmacy Ainsworth.

The wider and more important question is how Charles' personal views and interests will affect the royal family and the monarchy. His opinions, even if based on past actions, will still impact on the public mood and on policy. While the Queen was on the throne, the public attitude towards the monarchy and its role was unmovably positive. But now that era is well and truly over.

Apart from his love affair with quackery, Charles has, during the last year or so, provided plenty of reasons for unflattering headlines. Recent media coverage was dominated by a police investigation into allegations of cash for honours scandals linked to his charity, the Prince's Foundation.[311] There were also reports that Charles had accepted sizeable donations of plastic bags stuffed with cash from dubious Qatari sources.[312] Then there is the saga of his brother, Prince Andrew, and the convicted paedophile Jeffrey Epstein,[313] as well as the embarrassing revelations and allegations by his son, the Duke of Sussex.[314]

The public good will after his mother's death has seen the new Carolean age off to a strong start, but now Charles faces the formidable task to make it last. He has stepped into the Queen's pristine shoes only to discover that, within no time, they became covered in mud. Public support for him, his family, and the monarchy has become far less certain than it has been during his mother's 70-year reign.

A Renaissance man is a polymath and humanist with limitless capacities for development who is able to embrace all knowledge. Will King Charles grow into such an icon, or will he remain the ostentatious controversialist and quackery promoter that he has been so far? We shall see and, of course, we must wish him luck — at the very least, his reign is unlikely to be dull.[315]

Table 1. Important articles published since the first edition of this book. (Detailed discussions of these papers can be found on my blog—edzardernst.com.)

Title of article	Conclusion (quote) and reference
Homeopathic medicinal products for preventing and treating acute respiratory tract infections in children	[The evidence] did not show any consistent benefit of homeopathic medicinal products compared to placebo on ARTI recurrence or cure rates in children.[316]
The short-term effects of instrument-based mobilization compared with manual mobilization for low back pain: A randomized clinical trial	Both methods of lumbar spine mobilization demonstrated comparable improvements in pain and disability in patients with LBP, with neither method exhibiting superiority over the other.[317]
Real versus sham manual therapy in addition to therapeutic exercise in the treatment of non-specific shoulder pain: A randomized controlled trial	The addition of the manual therapy techniques applied in the present study to a therapeutic exercise protocol did not seem to add benefits to the management of subjects with non-specific shoulder pain.[318]
Bioenergy therapies as a complementary treatment: A systematic review to evaluate the efficacy of bioenergy therapies in relieving treatment toxicities in patients with cancer	Studies with high study quality could not find any difference between bioenergy therapies and active and passive control groups.[319]
Why homoeopathy is pseudoscience	Homoeopathy should be regarded as pseudoscience because its proponents claim scientific standing for it and produce argumentative bullshit to defend it, thus violating important epistemic standards central to science.[320]
Osteopathic manipulative treatment for pediatric conditions: An update of systematic review and meta-analysis	The effectiveness of OMT for selected pediatric populations remains unproven.[321]
Stenting as a treatment for craniocervical artery dissection: Improved major adverse cardiovascular event-free survival	Predisposing factors were noted in the majority (78%), including... chiropractic manipulation.[322]
Effectiveness and cost-effectiveness of universal school-based mindfulness training compared with normal school provision in reducing risk of mental	Findings do not support the superiority of SBMT over TAU in promoting mental health in adolescence.[323]

Title of article	Conclusion (quote) and reference
health problems and promoting well-being in adolescence: The MYRIAD cluster randomised controlled trial	
Effectiveness of a brief hypnotic induction in third molar extraction: A randomized controlled trial (HypMol)	A brief hypnotic induction with reduced preoperative local anesthetic use did not generally reduce post-treatment pain after third molar extraction more than regular local anesthetics.[324]
Integrative medicine during the intensive phase of chemotherapy in pediatric oncology in Germany: A randomized controlled trial with 5-year follow up	No beneficial effects of AST between group toxicity scores, overall or event-free survival were shown.[325]
Continued risk of dietary supplements adulterated with approved and unapproved drugs: Assessment of the US Food and Drug Administration's Tainted Supplements Database 2007 through 2021	The lack of disclosure of APIs in dietary supplements, circumventing the normal procedure with clinician oversight of prescription drug use, and the use of APIs that are banned by the Food and Drug Administration or used in combinations that were never studied are important health risks for consumers.[326]
Real-life drug-drug and herb-drug interactions in outpatients taking oral anticancer drugs: Comparison with databases	Potentially clinically relevant drug interaction were frequently identified in this study, showing that several databases and structured screening are required to detect more interactions and optimize medication safety.[327]
Homeopathy effects in patients during oncological treatment: A systematic review	For homeopathy, there is neither a scientifically based hypothesis of its mode of action nor conclusive evidence from clinical studies in cancer care.[328]
Does acupressure help reduce nausea and vomiting in palliative care patients? A double blind randomised controlled trial	Acupressure wristbands were no better than placebo for specialist palliative care in-patients with advanced cancer and nausea and vomiting.[329]
Homeopathy for COVID-19 in primary care: A randomized, double-blind, placebo-controlled trial (COVID-Simile study)	There was no statistically significant difference in the primary endpoints of Natrum muriaticum LM2 and placebo for mild COVID-19 cases.[330]
A randomized, blinded, placebo-controlled trial comparing antibody responses to homeopathic and conventional vaccines in university students	Homeopathic vaccines do not evoke antibody responses and produce a response that is similar to placebo.[331]
Hypericum perforatum to Improve Postoperative Pain Outcome After Monosegmental Spinal Sequestrectomy	No significant differences between the groups could be shown.[332]

Title of article	Conclusion (quote) and reference
(HYPOS): Results of a randomized, double-blind, placebo-controlled trial	
Efficacy of individualized homeopathic medicines in treatment of acne vulgaris: A double-blind, randomized, placebo-controlled trial	There was non-significant direction of effect favoring homeopathy against placebo in the treatment of AV.[333]
Assessing the magnitude of reporting bias in trials of homeopathy: A cross-sectional study and meta-analysis	Registration of published trials was infrequent, many registered trials were not published and primary outcomes were often altered or changed. This likely affects the validity of the body of evidence of homeopathic literature and may overestimate the true treatment effect of homeopathic remedies.[334]
Effects of flower essences on nursing students' stress symptoms: A randomized clinical trial	The intervention with flower essence therapy was not more effective than placebo in reducing stress signs and symptoms.[335]

Glossary

Acupuncture is an alternative therapy involving the insertion of needles into the skin and underlying tissues at acupuncture points for therapeutic or preventative purposes. Traditional acupuncture is mainly based on Taoist philosophy while Western acupuncture is based on neurophysiological concepts.

Adverse effects are unwanted side effects in layman's terms.

Anthroposophic medicine is an alternative medicine usually practised by doctors and based on the mystical concepts of Rudolf Steiner. Various treatments are employed by anthroposophic doctors, the most famous of which is mistletoe for cancer.

Aromatherapy is an alternative therapy that employs 'essential' oils usually combined with gentle massage; less commonly the oils are applied via inhalation.

Bias is the systematic deviation from the truth, generating conclusions that are misleading.

Chiropractic is an alternative therapy that was developed about 120 years ago by DD Palmer. The hallmark intervention of chiropractors is spinal manipulation which, they claim, is necessary to adjust 'subluxations'.

Chronic condition is a disease or symptom that has lasted for several, usually three or more, months.

Clinician is a healthcare professional quantifying the results of clinical trials or studies.

Cochrane Collaboration is an international organisation devoted to generating systematic reviews of high quality, transparency and integrity.

Compassion is the feeling that often arises when one is confronted with another person's suffering and one feels motivated to relieve that suffering.

Cost-effectiveness analysis is an economic evaluation where the relative costs and outcomes of two or more treatments for the same condition are compared.

Controlled clinical trial is a study where patients are divided into two or more groups receiving different interventions the effects of which are being compared at the end of the treatment period.

Critical thinking (critical analysis, assessment, or evaluation) is the process of conceptualising, applying, analysing, synthesising, and/or evaluating information gathered by observation, experience, reflection, reasoning, and/or communication.

Cure/curative treatment is a therapy that permanently frees a patient from a disease, as opposed to one that merely alleviates the symptoms.

Dietary supplement is a preparation intended to supplement the diet; it can contain all sorts of substances, e.g. vitamins, minerals, herbal remedies.

Detox is an umbrella term for numerous alternative therapies that allegedly eliminate toxins from the body.

Double-blind is the term used in clinical trials to indicate that both the patient as well as the investigators do not know whether the patient has been allocated to the control or the experimental group.

Effectiveness of a treatment refers to the clinical effects caused by the therapy (rather than by other phenomena such as the placebo effect) as demonstrated under real life conditions.

Efficacy of a treatment refers to its clinical effects caused by the therapy under strictly controlled conditions (some treatments are efficacious but not effective; for instance, they might have significant adverse effects which overshadow their clinical effects under real life conditions).

Empathy is the awareness of the feelings and emotions of other people. It is a key element of 'emotional intelligence' which connects oneself with others, because it is how we as individuals understand what others are experiencing as if we were feeling it ourselves.

Energy is the capacity to perform work and is measured in units of Joules. Energy exists in several forms such as heat, kinetic or mechanical energy, light, potential energy, electrical energy. In SCAM, the term is often applied to a patient's vital force as postulated by proponents of the long obsolete philosophy of vitalism.

Energy healing is an umbrella term for several alternative treatments that rely on the use of 'energy', i.e. vital force; examples are Reiki, Therapeutic Touch, and Johrei healing.

Evidence is the body of facts that leads to a given conclusion.

Evidence-based medicine (EBM) is the integration of best research evidence with clinical expertise and patient values. It thus rests on three pillars: external evidence, ideally from systematic reviews, the clinician's experience, and the patient's preferences.

Fallacy is a commonly used argument that appears to be logical but, in fact, is erroneous.

Herbal medicine (or phytotherapy) is the medicinal use of preparations that contain exclusively plant material.

Homeopathy is a therapeutic method using substances whose effects, when administered to healthy subjects, correspond to the manifestations of the disorder in the individual patient.

Hypothesis is a proposed explanation for a phenomenon.

Manual therapies are treatments performed by a therapist with her hands; examples include chiropractic, massage, osteopathy, shiatzu, or Bowen technique.

Medline is the world largest electronic database of medical articles; it contains millions of medical papers, including many on alternative medicine, and is freely accessible via the internet.

Mind–body therapies are alternative treatments which are thought to influence bodily functions via the mind.

National Health Service is the publicly funded healthcare system in the UK. Funded primarily by taxation, it provides healthcare to all legal residents of the UK. Specific policies vary between England, Scotland, Wales, and Northern Ireland.

Natural history of a disease describes the progress of a medical condition when left untreated.

Naturopathy is a type of healthcare which employs what nature provides (e.g. herbal extracts, manual therapies, heat and cold, water, and electricity) for stimulating the body's ability to heal itself.

Non-specific effects describe all phenomena which can determine the clinical outcome but are not due to the treatment *per se*. The best-known non-specific effect is the placebo effect.

Observational study is a non-experimental investigation, usually without a control group. In a typical observational study, patients receiving routine care are monitored as to the treatments administered and the outcomes observed.

Osteopathy is a manual therapy involving manipulation of the spine and other joints as well as mobilisation of soft tissues.

Pilot study is an investigation that is preliminary and typically aimed at determining whether a given protocol is feasible for testing a hypothesis.

Placebo is an inert treatment that has no effects *per se* but can appear to be effective through the placebo effect which essentially relies on conditioning and expectations.

Plausibility relates to the question whether there are logical explanations for an observed or postulated phenomenon. The plausibility of a therapy depends on whether its postulated mechanism of action is understandable in the light of established facts and science.

Post-marketing surveillance describes the monitoring of adverse effects of a therapy while it is used by millions of patients. In conventional medicine, it is usually achieved by a reporting scheme which notifies the regulator of all observed problems in clinical practice. In alternative medicine, no effective post-marketing surveillance systems are in place.

Potency is, according to homeopathic thinking, the 'power' of a remedy based on the degree to which it has been potentised, i.e. diluted and succussed (agitated). Low potency remedies are not highly diluted whereas high potency remedies are. Low potency remedies contain detectable concentrations of the starting material, whereas high potency remedies contain no detectable amount of the starting material.

Potentisation is the process of manufacturing a homeopathic remedy at a certain potency.

Pseudoscience is a fake that imitates real science without having all of its qualities.

Quality of life describes the state of well-being of a person. It can be measured by various means (e.g. validated questionnaires such as the 'SF36') and used to monitor the success of alternative therapies. It is often employed as an outcome measure in clinical trials.

Randomisation is a method used in controlled clinical trials for minimising bias; it involves dividing the total group of participants in typically two subgroups purely by chance, e.g. throwing dice. The effect of randomisation is that the two subgroups are comparable in all known and even unknown characteristics.

Randomised clinical trial (RCT) is a controlled clinical trial where patients are allocated to experimental or control groups by randomisation.

Reflexology is an alternative therapy employing manual pressure to specific areas of the body, usually the feet, which are claimed to correspond to internal organs with a view to generating positive health effects.

Regression towards the mean is the phenomenon that, over time, extreme values tend to move towards less extreme values. Patients normally consult clinicians when they are in an extreme situation (e.g. when they have much pain). Because of the regression towards the mean, they are likely to feel better the next time they see them. This change is regardless of the effects of any treatment they may have had. Regression towards the mean is therefore one of several phenomena that can make an ineffective therapy appear to be effective.

Sceptic/ skeptic is a person who habitually doubts notions which most other people view as established.

Science can be defined as the identification, description, observation, experimental investigation, and theoretical explanation of phenomena (see also pseudoscience).

Spinal manipulation is the term used for the manual adjustments of subluxations of the vertebrae often employed by chiropractors and osteopaths.

Subluxation, as used by chiropractors, is an abnormality in the relative position of vertebrae which chiropractors claim to be able to adjust.

Symptomatic treatment is a treatment that alleviates symptoms without necessarily treating the cause of a condition.

Systematic review is an evaluation of the totality of the available evidence related to a specific research question. Systematic reviews minimise the bias inherent in each single study. If they include a mathematical pooling of data from single studies into a new overall result, they are called meta-analyses.

Theory is the result of abstract thinking about generalised explanations of how nature works. A theory provides an explanatory framework for a set of observations. From the assumptions of the explanation follow several possible hypotheses that can be tested in order to provide evidence for or against the theory.

Traditional Chinese Medicine (TCM) is a diagnostic and therapeutic system based on the Taoist philosophy of yin and yang. It includes alternative treatments that emerged from China, including acupuncture, herbal medicine, tui-na (Chinese massage), tai chi, and diet.

Vital energy is a metaphysical concept of a power that allegedly animates all organisms.

Vitalism is the metaphysical concept that life depends on a vital energy or force distinct from chemical, physical, or other principles. It is a concept found in many forms of alternative medicine, e.g. *chi* in China, *pneuma* in ancient Greece, and *prana* in India. The common denominator is the assumption that a metaphysical energy animates all living systems.

End Notes

1. https://www.princeofwales.gov.uk/speech/article-hrh-prince-wales-titled-science-and-homeopathy-must-work-harmony-daily-telegraph
2. https://www.amazon.co.uk/Harmony-New-Way-Looking-World/dp/0007348037
3. https://www.amazon.co.uk/Charles-Heart-King-Catherine-Mayer-ebook/dp/B00O72S27U
4. Ioannidis, J.P.A., Boyack, K.W. & Baas, J. (2020) Updated science-wide author databases of standardized citation indicators, *PLoS Biology*, 18 (10), e3000918.
5. https://www.amazon.co.uk/Scientist-Wonderland-Searching-Finding-Trouble/dp/1845407776
6. https://www.amazon.co.uk/Prince-Wales-Biography-Jonathan-Dimbleby/dp/068812996X
7. Wieland, L.S., Manheimer, E. & Berman, B.M. (2011) Development and classification of an operational definition of complementary and alternative medicine for the Cochrane collaboration, *Alternative Therapies in Health & Medicine*, 17 (2), pp. 50–59.
8. https://www.hive.co.uk/Product/Edzard-Ernst/Homeopathy---The-Undiluted-Facts--Including-a-Comprehensive-A-Z-Lexicon/19719982
9. Hewitt, D. & Wood, P.H. (1975) Heterodox practitioners and the availability of specialist advice, *Rheumatology & Rehabilitation*, 14 (3), pp. 191–199.
10. Ernst, E. (2013) Thirteen follies and fallacies about alternative medicine, *EMBO Reports*, 14 (12), pp. 1025–1026.
11. https://www.amazon.co.uk/SCAM-So-Called-Alternative-Medicine-Societas/dp/1845409701
12. https://www.amazon.co.uk/Charles-Misunderstood-Prince-biography-everyones-ebook/dp/B01MV3HBG6
13. https://www.amazon.co.uk/gp/product/0719555809
14. https://www.sa-venues.com/things-to-do/freestate/visit-the-laurens-van-der-post-memorial-garden/
15. https://www.theguardian.com/books/2001/sep/22/biography.artsandhumanities
16. https://www.theguardian.com/world/2001/feb/04/uk.vanessathorpe
17. https://www.nytimes.com/2002/08/03/books/master-storyteller-or-master-deceiver.html

18 https://slate.com/news-and-politics/2010/06/prince-charles-sinister-speech-attacks-science-and-good-sense.html
19 https://www.theguardian.com/uk/2005/aug/25/monarchy.health
20 https://www.amazon.co.uk/Rebel-Prince-Defiance-explosive-biography-ebook/dp/B079172F2M/
21 https://www.amazon.co.uk/Radical-Prince-Practical-Vision-Wales/dp/086315431X
22 https://www.amazon.co.uk/Alternative-Therapy-Board-Science-Education/dp/B001A0HIJ0
23 https://www.amazon.co.uk/Complementary-Medicine-Approaches-Good-Practice/dp/0192861662
24 https://www.rsm.ac.uk/about-us/our-mission/
25 https://www.independent.co.uk/news/obituaries/surgeon-vice-admiral-sir-james-watt-doctor-who-promoted-christian-values-royal-navy-1936039.html
26 https://www.amazon.co.uk/Talking-Health-Conventional-Complementary-Approaches/dp/0905958640
27 Still, A.T. (1899) *Philosophy of Osteopathy*, Ann Arbor: Edward Brothers Inc.
28 https://www.osteopathy.org.uk/visiting-an-osteopath/about-osteopathy/
29 https://www.amazon.co.uk/Alternative-Medicine-Critical-Assessment-Modalities-ebook/dp/B07TS1QXX6
30 https://oialliance.org/wp-content/uploads/2014/01/OIA-Stage-2-Report.pdf
31 https://www.princeofwales.gov.uk/speeches?pow_check=on&keywords=integrative+medicine&from_date=&to_date=#na
32 https://appletzara.wordpress.com/2016/04/24/osteopathy-part-2-a-review-of-100-osteopathy-websites/
33 Licciardone, J.C., Brimhall, A.K. & King, L.N. (2005) Osteopathic manipulative treatment for low back pain: A systematic review and meta-analysis of randomized controlled trials, *BMC Musculoskeletal Disorders*, 6, 43.
34 Posadzki, P. & Ernst, E. (2011) Osteopathy for musculoskeletal pain patients: A systematic review of randomized controlled trials, *Clinical Rheumatology*, 30 (2), pp. 285–291.
35 Verhaeghe, N., Schepers, J., van Dun, P. & Annemans, L. (2018) Osteopathic care for spinal complaints: A systematic literature review, *PLoS One*, 13 (11), e0206284. Erratum in: *PLoS One* (2019) 14 (8), e0221140.
36 Posadzki, P., Lee, M.S. & Ernst, E. (2013) Osteopathic manipulative treatment for pediatric conditions: A systematic review, *Pediatrics*, 132 (1), pp. 140–152. doi: 10.1542/peds.2012-3959.
37 https://www.timeshighereducation.com/news/british-school-of-osteopathy-wins-university-college-status
38 https://edzardernst.com/2016/03/nice-no-longer-recommends-acupuncture-chiropractic-or-osteopathy-for-low-back-pain/
39 Jonas, C. (2018) Musculoskeletal therapies: Osteopathic manipulative

treatment, *FP Essentials*, 470, pp. 11–15.
40. Data based on Medline search conducted in April 2021.
41. https://www.amazon.co.uk/Chiropractic-Not-All-That-Cracked/dp/3030531171
42. https://www.aecc.ac.uk/about/who-we-are/our-history/
43. https://www.gcc-uk.org/assets/publications/The_Chiropractors_Act_1994_and_opening_of_the_GCC_register_1999.pdf
44. https://www.ebm-first.com/chiropractic/uk-chiropractic-issues/2192-the-pursuit-for-chiropractic-legislation.html
45. https://www.ebm-first.com/chiropractic/uk-chiropractic-issues/2192-the-pursuit-for-chiropractic-legislation.html
46. https://www.dailymail.co.uk/news/article-1232390/Prince-Charles-calls-alternative-medicine-formally-regulated.html
47. https://www.telegraph.co.uk/news/health/news/11500275/Herbal-doctors-will-not-be-regulated-despite-pleas-from-Prince-Charles.html
48. https://rcc-uk.org/rcc-history/
49. Mirtz, T.A., Morgan, L., Wyatt, L.H. & Greene, L. (2009) An epidemiological examination of the 1961 subluxation construct using Hill's criteria of causation, *Chiropractic & Osteopathy*, 2 (17), 13.
50. Ernst, E. (2008) Chiropractic: A critical evaluation, *Journal of Pain & Symptom Management*, 35 (5), pp. 544–562.
51. Ernst, E. (2009) Chiropractic maintenance treatment, a useful preventative approach?, *Preventative Medicine*, 49 (2–3), pp. 99–100.
52. Stevinson, C. & Ernst, E. (2002) Risks associated with spinal manipulation, *American Journal of Medicine*, 112 (7), pp. 566–571.
53. https://rcc-uk.org/wp-content/uploads/2018/04/Chiropractic_the-facts_v3.pdf
54. https://edzardernst.com/2019/11/we-hope-that-the-publicity-surrounding-this-event-will-highlight-the-dangers-of-chiropractic-a-statement-of-the-family-of-the-man-who-died-after-treatment-of-a-vertebral-subluxation-complex/
55. https://www.padbergcorrigan.com/malpractice-neck-manipulation-stroke/
56. https://en.wikipedia.org/wiki/British_Chiropractic_Association_v_Singh
57. https://edzardernst.com/2015/02/the-uk-general-chiropractic-council-fit-for-purpose/
58. https://en.wikipedia.org/wiki/The_Prince's_Foundation_for_Integrated_Health
59. https://www.theguardian.com/uk/2010/apr/26/prince-charles-aide-homeopathy-charity-arrested
60. https://www.princeofwales.gov.uk/speech/article-hrh-prince-wales-titled-landmarks-development-integrated-medicine-nhs-magazine
61. https://www.princeofwales.gov.uk/speech/article-hrh-prince-wales-titled-landmarks-development-integrated-medicine-nhs-magazine
62. https://www.amazon.co.uk/Suckers-Alternative-Medicine-Makes-

Fools/dp/0099522861
[63] https://www.theguardian.com/uk/2004/dec/23/health.monarchy
[64] https://wikimili.com/en/The_Prince's_Foundation_for_Integrated_Health#cite_note-charity-comm-13
[65] Ernst, E. (2005) Consumer guides for complementary medicine, *AIDS & Hepatitis Digest*, 108, pp. 5-6.
[66] https://sciencebasedmedicine.org/dr-michael-dixon-a-pyromaniac-in-a-field-of-integrative-straw-men/
[67] http://news.bbc.co.uk/2/hi/uk_news/8654679.stm
[68] https://www.quackometer.net/blog/2010/04/prince-of-wales-charity-faces-imminent-closure.html
[69] https://www.independent.co.uk/voices/commentators/edzard-ernst-why-alternative-medicine-wins-from-the-foundation-s-demise-1959615.html
[70] https://www.amazon.co.uk/Time-Heal-Tales-Country-Doctor/dp/1913491161
[71] https://chiro.org/alt_med_abstracts/FULL/Prince_Charles_Alternative_Medicine.shtml
[72] House of Lords Select Committee on Science and Technology (2000) *Complementary and Alternative Medicine. Report. 6th report. Session 1999-2000.* (HL123) ISBN 0 10 483 1006. London: The Stationery Office.
[73] Ernst, E. & White, A. (2000) The BBC survey of complementary medicine use in the UK, *Complementary Therapies in Medicine*, 8 (1), pp. 32-36.
[74] Astin, J.A. (1998) Why patients use alternative medicine: Results of a national study, *JAMA*, 279 (19), pp. 1548-53.
[75] Hunt, K.J., Coelho, H.F., Wider, B., Perry, R., Hung, S.K., Terry, R. & Ernst, E. (2010) Complementary and alternative medicine use in England: Results from a national survey, *International Journal of Clinical Practice*, 64 (11), pp. 1496-1502.
[76] Ernst, E., Posadzki, P. & Lee, M.S. (2011) Reflexology: An update of a systematic review of randomised clinical trials, *Maturitas*, 68 (2), pp. 116-120.
[77] https://edzardernst.com/2016/12/the-mainstreaming-of-quackery-the-role-of-the-nccih/
[78] Ernst, E. & Wider, B. (2002) Medical research charities should fund more trials, *BMJ*, 325 (7374), 1245.
[79] Ernst, E. (2006) Prevalence surveys: To be taken with a pinch of salt, *Complementary Therapies in Clinical Practice*, 12 (4), pp. 272-275.
[80] Brennan, Z. & Hellen, N. (2001) Charles Helps to Build 'New Age' Hospital, *The Sunday Times*, 26 August.
[81] http://www.ntskeptics.org/news/news2001-09-02.htm
[82] https://edzardernst.com/2015/01/rudolf-hess-hitlers-deputy-on-alternative-medicine/
[83] https://www.kwakzalverij.nl/behandelwijzen/homeopathie/der-donner-bericht/

[84] Ernst, E. (2001) 'Neue Deutsche Heilkunde': Complementary/alternative medicine in the Third Reich, *Complementary Therapies in Medicine*, 9 (1), pp. 49–51.
[85] http://www.integratedmed.co.uk/dr-mosaraf-ali/
[86] https://www.wellbuzz.com/dr-ozs-advice/dr-oz-dr-mosaraf-ali-the-controversial-healer/
[87] https://www.mirror.co.uk/lifestyle/sex-relationships/stay-safe-if-you-go-alternative-441183
[88] http://arhiva.nacional.hr/en/clanak/26534/dr-ali-physician-to-the-british-royal-family
[89] https://www.amazon.co.uk/Integrated-Health-Bible-Well-Being-22-Feb-2001/dp/B011T7P6R4
[90] https://www.theguardian.com/society/2008/sep/07/health
[91] Fox, M., Dickens, A., Greaves, C., Dixon, M. & James, M. (2006) Marma therapy for stroke rehabilitation – a pilot study, *Journal of Rehabilitation Medicine*, 38 (4), pp. 268–271.
[92] https://www.9news.com.au/entertainment/amputee-sues-prince-charles-doctor/df129fd3-03a4-4c9a-bbed-66225a6317ed
[93] https://themachoresponse.blogspot.com/2008/09/charles-and-camilla-cant-cure-you.html
[94] https://www.bmj.com/content/322/7279/181.1
[95] HRH Prince Charles (2012) Integrated health and post-modern medicine, *Journal of the Royal Society of Medicine*, 105 (12), pp. 496–498.
[96] HRH Prince Charles (2021) A message from HRH The Prince of Wales, honorary fellow of the Royal College of Physicians, *Future Healthcare Journal*, 8 (1), pp. 5–7.
[97] https://edzardernst.com/2015/01/rudolf-hess-hitlers-deputy-on-alternative-medicine/
[98] https://edzardernst.com/2017/03/integrative-medicine-physicians-tend-to-harbour-anti-vaccination-views/
[99] https://www.amazon.co.uk/Alternative-Medicine-Critical-Assessment-Modalities-ebook/dp/B07TS1QXX6/ref=sr_1_5
[100] https://edzardernst.com/2021/01/a-new-definition-of-integrated-medicine/
[101] https://respectfulinsolence.com/2012/02/28/boiling-integrative-medicine-down-to-its/
[102] https://edzardernst.com/2015/07/integratedintegrative-medicine-a-paradise-for-charlatans/
[103] https://edzardernst.com/2016/05/integrative-medicine-one-of-the-most-colossal-deceptions-in-healthcare-today/
[104] https://gerson.org/gerpress/
[105] http://news.bbc.co.uk/2/hi/health/3876431.stm
[106] Chabot, J.A., Tsai, W.Y., Fine, R.L., Chen, C., Kumah, C.K., Antman, K.A. & Grann, V.R. (2010) Pancreatic proteolytic enzyme therapy compared with gemcitabine-based chemotherapy for the treatment of pancreatic cancer,

Journal of Clinical Oncology, 28 (12), pp. 2058–2063.
107 https://www.cancerresearchuk.org/about-cancer/cancer-in-general/treatment/complementary-alternative-therapies/individual-therapies/gerson
108 https://www.amazon.co.uk/Cancer-Therapy-Results-Fifty-Reprint/dp/1684222567
109 https://www.nhs.uk/conditions/herbal-medicines/
110 Ernst, E. (2007) Herbal medicine: Buy one, get two free, *Postgraduate Medical Journal*, 83 (984), pp. 615–616.
111 https://www.amazon.co.uk/Rational-Phytotherapy-Reference-Physicians-Pharmacists/dp/3642074065
112 https://www.bmj.com/content/350/bmj.h2642
113 https://www.bbc.com/news/uk-32726099
114 http://www.dcscience.net/2015/05/15/prince-charles-letters-confirm-that-hes-not-fit-to-be-king/
115 https://whyweprotest.net/threads/king-charles.116325/
116 https://www.theguardian.com/uk-news/2015/jun/04/black-spider-memos-prince-charles-lobbied-homeopathy-funding-nhs
117 Hung, S.K. & Ernst, E. (2010) Herbal medicine: An overview of the literature from three decades, *Journal of Dietary Supplements*, 7 (3), pp. 217–226.
118 Guo, R., Canter, P.H. & Ernst, E. (2007) A systematic review of randomised clinical trials of individualised herbal medicine in any indication, *Postgraduate Medical Journal*, 83 (984), pp. 633–637.
119 Lechner, M., Steirer, I., Brinkhaus, B., Chen, Y., Krist-Dungl, C., Koschier, A., Gantschacher, M., Neumann, K. & Zauner-Dungl, A. (2011) Efficacy of individualized Chinese herbal medication in osteoarthrosis of hip and knee: A double-blind, randomized-controlled clinical study, *Journal of Alternative & Complementary Medicine*, 17 (6), pp. 539–547.
120 Cohen, P.A. & Ernst, E. (2010) Safety of herbal supplements: A guide for cardiologists, *Cardiovascular Therapeutics*, 28 (4), pp. 246–253.
121 Posadzki, P., Watson, L. & Ernst, E. (2013) Contamination and adulteration of herbal medicinal products (HMPs): An overview of systematic reviews, *European Journal of Clinical Pharmacology*, 69 (3), pp. 295–307.
122 http://news.bbc.co.uk/2/hi/health/8388985.stm
123 https://nimh.org.uk/media-centre/
124 Smallwood, C. (2005) The role of complementary and alternative medicine in the NHS. An investigation into the potential contribution of mainstream complementary therapies to healthcare in the UK, [Online], http://www.freshminds.co.uk/ PDF/THE%20REPORT.pdf
125 https://www.freshminds.co.uk/
126 http://news.bbc.co.uk/2/hi/health/4312780.stm
127 Thompson Coon, J. & Ernst, E. (2005) A systematic review of the economic evaluation of complementary and alternative medicine, *Perfusion*, 18, pp. 202–214.
128 Henderson, M. (2008) Prince of Wales's guide to alternative medicine

'inaccurate' – Times Online, *The Times*, 17 April.
[129] https://en.wikipedia.org/wiki/Smallwood_Report
[130] Fenton, B. (2005) Awkward moment for prince at the chemist's, *Daily Telegraph*, 13 October, [Online], http://www.telegraph.co.uk/news/main.jhtml?xml=/news/2005/10/13/nprin13.xml
[131] Horton, R. (2005) Rational medicine is being undermined, *The Guardian*, 8 October, [Online], http://www.guardian.co.uk/letters/story/0,,1587525,00.html
[132] Assendelft, W.J.J., Morton, S.C., Yu, E.I., et al. (2004) Spinal manipulative therapy for low-backpain, *Cochrane Database Systematic Review*, 1, art. CD000447.
[133] Turner, R.B., Bauer, R., Woelkart, K., et al. (2005) An evaluation of Echinacea angustifolia in experimental rhinovirus infections, *New England Journal of Medicine*, 353, pp. 341–348.
[134] McCarney, R.W., Linde, K. & Lasserson, T.J. (2004) Homeopathy for chronic asthma, *Cochrane Database Systematic Review*, 1, art. CD000353.
[135] For example, Shang, A., Huwiler-Muntener, K., Nartey, L., et al. (2005) Are the clinical effects of homoeopathy placebo effects? Comparative study of placebo-controlled trials of homoeopathy and allopathy, *Lancet*, 366, pp. 726–732.
[136] Canter, P.H., Thompson Coon, J. & Ernst, E. (2005) Cost effectiveness of complementary treatments in the United Kingdom: Systematic review, *BMJ*, 331, p. 881.
[137] https://www.theguardian.com/books/2007/mar/24/politicalbooks.biography
[138] https://www.quackometer.net/blog/2010/04/dame-shirley-porter-funded-prince-charles-political-report-on-nhs-alternative-medicine.html
[139] https://www.thetimes.co.uk/article/may-2006-doctors-campaign-against-alternative-therapies-lrr265r2kcg
[140] https://edzardernst.com/2019/08/blowing-my-own-trumpet-to-the-tune-of-a-standardized-citation-metrics-author-database-annotated-for-scientific-field/
[141] https://www.princeofwales.gov.uk/speech/speech-hrh-prince-wales-integrated-healthcare-world-health-assembly-geneva-switzerland
[142] https://fallacyinlogic.com/appeal-to-tradition-fallacy-definition-and-examples/
[143] Gaster, B. & Holroyd, J. (2000) St John's wort for depression: A systematic review, *Archives of Internal Medicine*, 160 (2), pp. 152–156.
[144] Ernst, E., Lee, M.S. & Choi, T.Y. (2011) Acupuncture: Does it alleviate pain and are there serious risks? A review of reviews, *Pain*, 152 (4), pp. 755–764.
[145] Kwon, Y.D., Pittler, M.H. & Ernst, E. (2006) Acupuncture for peripheral joint osteoarthritis: A systematic review and meta-analysis, *Rheumatology (Oxford)*, 45 (11), pp. 1331–1337.
[146] Ezzo, J., Vickers, A., Richardson, M.A., Allen, C., Dibble, S.L., Issell, B., Lao, L., Pearl, M., Ramirez, G., Roscoe, J.A., Shen, J., Shivnan, J.,

Streitberger, K., Treish, I. & Zhang, G. (2005) Acupuncture-point stimulation for chemotherapy-induced nausea and vomiting, *Journal of Clinical Oncology*, 23 (28), pp. 7188–7198.

[147] Ernst, E. (1999) Second thoughts about safety of St John's wort, *Lancet*, 354 (9195), pp. 2014–2016. Erratum in: *Lancet* (2000) 355 (9203), p. 580.

[148] https://edzardernst.com/2018/06/first-do-no-harm-what-does-it-mean-how-does-it-apply-to-alternative-medicine/

[149] Graziose, R., Lila, M.A. & Raskin, I. (2010) Merging traditional Chinese medicine with modern drug discovery technologies to find novel drugs and functional foods, *Current Drug Discovery Technologies*, 7 (1), pp. 2–12.

[150] https://edzardernst.com/2015/10/a-nobel-prize-for-tcm/

[151] https://www.ebm-first.com/a-close-look-at-alternative-medicine/nhs-told-to-abandon-cam.html

[152] https://apps.who.int/gb/ebwha/pdf_files/EB134/B134_24-en.pdf

[153] Gillon, R. (1985) 'Primum non nocere' and the principle of non-maleficence, *British Medical Journal (Clinical Research Edition)*, 291 (6488), pp. 130–131.

[154] https://www.thetimes.co.uk/article/prince-charles-is-talking-like-a-crank-on-covid-19-jzxgrxw28

[155] https://edzardernst.com/2018/05/china-power-and-influence/

[156] https://media.nature.com/original/magazine-assets/d41586-017-07650-6/d41586-017-07650-6.pdf

[157] https://www.caixinglobal.com/2020-12-02/beijing-drops-plan-to-criminalize-criticism-of-chinese-medicine-101634925.html

[158] https://sciencebasedmedicine.org/what-is-traditional-chinese-medicine/

[159] Melchart, D., Hager, S., Albrecht, S., Dai, J., Weidenhammer, W. & Teschke, R. (2017) Herbal Traditional Chinese Medicine and suspected liver injury: A prospective study, *World Journal of Hepatology*, 9 (29), pp. 1141–1157.

[160] Izzo, A.A. & Ernst, E. (2009) Interactions between herbal medicines and prescribed drugs: An updated systematic review, *Drugs*, 69 (13), pp. 1777–1798.

[161] https://edzardernst.com/2016/07/promoting-rhino-horn-as-medicine-at-an-australian-university-has-this-contributed-to-the-exponential-rise-in-rhino-poaching/

[162] https://blogs.sciencemag.org/pipeline/archives/2020/05/12/more-chinese-traditional-medicine-unfortunately

[163] https://easac.eu/publications/details/traditional-chinese-medicine-a-statement-by-easac-and-feam

[164] Tang, J.L., Zhan, S.Y. & Ernst, E. (1999) Review of randomised controlled trials of traditional Chinese medicine, *BMJ*, 319 (7203), pp. 160–161.

[165] https://edzardernst.com/2016/10/data-fabrication-in-china-is-an-open-secret/

[166] https://www.rfa.org/english/news/china/clinical-fakes-09272016141438.html

[167] https://www.bbc.com/news/uk-politics-28066081
[168] https://www.health-ni.gov.uk/
[169] https://www.quackometer.net/blog/2009/02/northern-ireland-nhs-alternative.html
[170] http://www.dcscience.net/2007/02/09/peter-hain-and-getwelluk-pseudoscience-and-privatisation-in-northern-ireland/
[171] https://www.growthbusiness.co.uk/social-entrepreneur-boo-armstrong-1020162/
[172] https://einsteinmed.org/departments/family-social-medicine/bravenet/about/collaborative/
[173] https://www.quackometer.net/blog/2008/06/bravewell-and-prince.html
[174] http://www.fourwinds10.com/siterun_data/health/holistic_alternative_medicine/news.php?q=1195513509
[175] https://www.forbes.com/2006/11/17/alternative-medicine-christy-mack-biz-cz_rl_1120bravewell/?sh=778aa9a06289
[176] https://sciencebasedmedicine.org/bravewell-collaborative-maps-the-state-of-integrative-medicine/
[177] https://bravewell.org/bravewell_collaborative/president_letter/
[178] https://en.wikipedia.org/wiki/Waitrose_Duchy_Organic
[179] https://www.princeofwales.gov.uk/what-duchy-originals-it-anything-do-duchy-cornwall
[180] https://en.wikipedia.org/wiki/Nelsons_(Homeopathy)
[181] https://www.theguardian.com/uk/2009/mar/11/prince-charles-detox-tincture
[182] Pizzorno, J.E. & Murray, M.T. (1999) *Textbook of Natural Medicine*, London: Churchill Livingstone.
[183] Ernst, E. (2012) Alternative detox, *British Medical Bulletin*, 101, pp. 33–38.
[184] https://www.telegraph.co.uk/news/uknews/theroyalfamily/5024341/Prince-Charles-Duchy-Originals-ordered-to-remove-misleading-herbal-remedy-claims.html
[185] https://cargo-cult-science.blogspot.com/2009/03/duchy-originals-tinctures-never.html
[186] https://theness.com/neurologicablog/index.php/duchy-originals-detox-tincture/
[187] https://www.msn.com/en-gb/health/other/detox-tincture-qanda/ar-AA2e3MQ
[188] https://freethinkeruk.wordpress.com/tag/duchy-originals/
[189] https://publiclawforeveryone.com/2015/03/26/of-black-spiders-and-constitutional-bedrock-the-supreme-courts-judgment-in-evans/
[190] https://www.mirror.co.uk/news/uk-news/prince-charles-letters-royal-wrote-5823182
[191] https://www.amazon.co.uk/Desktop-Guide-Complementary-Alternative-Medicine/dp/0723432074
[192] Ernst, E. (2002) A systematic review of systematic reviews of homeopathy, *British Journal of Clinical Pharmacology*, 54 (6), pp. 577–582.

193 https://www.amazon.co.uk/More-Harm-than-Good-Complementary-ebook/dp/B078ZQXQNP
194 Canter, P.H., Coon, J.T. & Ernst, E. (2006) Cost-effectiveness of complementary therapies in the United kingdom—a systematic review, *Evidence-Based Complementary & Alternative Medicine*, 3 (4), pp. 425–432.
195 https://www.independent.co.uk/news/people/prince-charles-saviour-nation-new-book-highlights-concerns-about-how-political-he-will-be-when-he-eventually-becomes-king-10016111.html
196 https://www.princeofwales.gov.uk/royal-duties
197 https://www.independent.co.uk/life-style/health-and-families/health-news/he-s-it-again-prince-charles-accused-lobbying-health-secretary-over-homeopathy-8723145.html
198 https://www.theguardian.com/commentisfree/2015/may/13/guardian-view-prince-charles-black-spider-letters
199 https://collegeofmedicine.org.uk/visionandvalues/
200 https://collegeofmedicine.org.uk/hrh-the-prince-of-wales-is-announced-as-college-of-medicine-patron/
201 See, for example, https://www.bmj.com/content/341/bmj.c6126.full, https://www.theguardian.com/lifeandstyle/2010/aug/02/prince-charles-college-medicine-holistic-complementary, https://www.bmj.com/content/342/bmj.d3712.full
202 http://www.dcscience.net/2010/10/29/dont-be-deceived-the-new-college-of-medicine-is-a-fraud-and-delusion/
203 https://timesofindia.indiatimes.com/city/bengaluru/healing-the-world-the-alternative-way/articleshow/25780194.cms
204 https://en.wikipedia.org/wiki/College_of_Medicine_(UK)
205 https://collegeofmedicine.org.uk/our-constitution/
206 https://collegeofmedicine.org.uk/postgraduate-diploma-integrative-medicine/
207 https://crossfieldsinstitute.com/steiner-early-years-level-4/
208 https://edzardernst.com/2016/08/the-only-accredited-integrative-medicine-diploma-currently-available-in-the-uk/
209 http://skepdic.com/neurolin.html
210 Sturt, J., Ali, S., Robertson, W., Metcalfe, D., Grove, A., Bourne, C. & Bridle, C. (2012) Neurolinguistic programming: A systematic review of the effects on health outcomes, *British Journal of General Practice*, 62 (604), e757–64.
211 https://edzardernst.com/2020/02/thought-field-therapy-a-couse-offered-by-the-college-of-quack-medicine/
212 https://en.wikipedia.org/wiki/List_of_psychotherapies
213 https://selfcaretoolkit.net/
214 https://ourhealth.directory/
215 Hansen, J.P., Pareek, M., Hvolby, A., Schmedes, A., Toft, T., Dahl, E. & Nielsen, C.T. (2019) Vitamin D3 supplementation and treatment outcomes in patients with depression (D3-vit-dep), *BMC Research Notes*, 12 (1), 203.
216 von Berens, Å., Fielding, R.A., Gustafsson, T., Kirn, D., Laussen, J., Nydahl,

M., Reid, K., Travison, T.G., Zhu, H., Cederholm, T. & Koochek, A. (2018) Effect of exercise and nutritional supplementation on health-related quality of life and mood in older adults: The VIVE2 randomized controlled trial, *BMC Geriatrics*, 18 (1), 286.

[217] Sahraian, A., Ghanizadeh, A. & Kazemeini, F. (2015) Vitamin C as an adjuvant for treating major depressive disorder and suicidal behavior, a randomized placebo-controlled clinical trial, *Trials*, 16, 94.

[218] https://collegeofmedicine.org.uk/ten-years-of-the-college-of-medicine-we-celebrate-a-decade-of-huge-achievement/

[219] https://www.telegraph.co.uk/news/uknews/theroyalfamily/7147870/Prince-of-Wales-I-was-accused-of-being-enemy-of-the-Enlightenment.html

[220] https://theconversation.com/quacktitioner-royal-is-a-menace-to-the-constitution-and-public-health-16448

[221] http://www.dcscience.net/2013/07/30/the-quacktitioner-royal-is-a-threat-to-constitutional-government-and-to-the-health-of-the-nation/

[222] https://en.wikipedia.org/wiki/Post_hoc_ergo_propter_hoc

[223] https://web.archive.org/web/20110723060336/http://www.crhp.net/article1.html

[224] Shepherd, S. & Kay, A.C. (2012) On the perpetuation of ignorance: System dependence, system justification, and the motivated avoidance of sociopolitical information, *Journal of Personality & Social Psychology*, 102 (2), pp. 264–280.

[225] Skrabanek, P. (1984) Acupuncture and the age of unreason, *Lancet*, 1 (8387), pp. 1169–1171.

[226] https://www.spectator.co.uk/article/the-madness-of-charles-iii

[227] Coates, J.R. & Jobst, K.A. (1998) Integrated healthcare: A way forward for the next five years? A discussion document from the Prince of Wales's Initiative on Integrated Medicine, *Journal of Alternative & Complementary Medicine*, 4 (2), pp. 209–247. Erratum in: *Journal of Alternative & Complementary Medicine* (1998), 4 (3), p. 353.

[228] https://www.tonyjuniper.com/

[229] https://en.wikipedia.org/wiki/Ian_Skelly

[230] https://www.amazon.co.uk/Harmony-New-Way-Looking-World/dp/0007348037

[231] https://www.amazon.co.uk/Prince-Charles-Passions-Paradoxes-Improbable/dp/1432847619

[232] https://www.spectator.co.uk/article/the-madness-of-charles-iii

[233] https://www.quackometer.net/blog/2010/12/the-new-age-medicine-of-prince-charles.html

[234] Ernst, E. (2009) Acupuncture: What does the most reliable evidence tell us?, *Journal of Pain & Symptom Management*, 37 (4), pp. 709–714.

[235] Ernst, E. (1999) Iridology: A systematic review, *Forsch Komplementarmed*, 6 (1), pp. 7–9.

[236] https://www.aerzteblatt.de/nachrichten/97971/Augenaerzte-warnen-vor-sogenannter-Irisdiagnostik

[237] Liu, Q., Yue, X.Q. & Ling, C.Q. (2003) [Researches into the modernization of tongue diagnosis: In retrospect and prospect], *Zhong Xi Yi Jie He Xue Bao*,1 (1), pp. 66–70. Chinese.
[238] Yue, X.Q. & Liu, Q. (2004) [Analysis of studies on pattern recognition of tongue image in traditional Chinese medicine by computer technology], *Zhong Xi Yi Jie He Xue Bao*, 2 (5), pp. 326–329. Chinese.
[239] Wei, B.G., Shen, L.S., Wang, Y.Q., Wang, Y.G., Wang, A.M. & Zhao, Z.X. (2002) [A digital tongue image analysis instrument for Traditional Chinese Medicine], *Zhongguo Yi Liao Qi Xie Za Zhi*, 26 (3), pp. 164–166, 169. Chinese.
[240] Lee, J.A., Ko, M.M., Kang, B.K., Alraek, T., Birch, S. & Lee, M.S. (2014) Interobserver reliability of four diagnostic methods using traditional Korean medicine for stroke patients, *Evidence Based Complementary & Alternative Medicine*, 465471.
[241] Ernst, E., Posadzki, P. & Lee, M.S. (2011) Reflexology: An update of a systematic review of randomised clinical trials, *Maturitas*, 68 (2), pp. 116–120.
[242] Williamson, J., White, A., Hart, A. & Ernst, E. (2002) Randomised controlled trial of reflexology for menopausal symptoms, *BJOG*, 109 (9), pp. 1050–1055.
[243] http://safehomediy.com/prince-of-wales-uses-homeopathic-treatments-for-his-livestock/
[244] https://www.dailymail.co.uk/news/article-3587406/Prince-Charles-uses-homeopathy-animals-organic-farms-reduce-reliance-antibiotics.html
[245] Sharma, H., Chandola, H.M., Singh, G. & Basisht, G. (2007) Utilization of Ayurveda in health care: An approach for prevention, health promotion, and treatment of disease. Part 1 — Ayurveda, the science of life, *Journal of Alternative & Complementary Medicine*, 13 (9), pp. 1011–1019.
[246] https://www.newindianexpress.com/nation/2018/apr/18/prime-minister-modi-arrives-in-uk-for-bilateral-meetings-commonwealth-heads-of-government-meeting-1803093.html
[247] https://www.express.co.uk/news/royal/948053/prince-charles-science-museum-plaque-india-prime-minister-narendra-modi
[248] https://collegeofmedicine.org.uk/college-of-medicine-and-indian-government-to-join-forces-to-create-ayush-centres-of-excellence-in-uk/
[249] https://presidentofindia.nic.in/press-release-detail.htm?1731
[250] https://edzardernst.com/2019/03/yoga-and-the-nhs/
[251] https://www.theguardian.com/uk-news/2021/may/28/prince-charles-advises-people-recovering-from-covid-to-practise-yoga
[252] Ernst, E. (2002) Heavy metals in traditional Indian remedies, *European Journal of Clinical Pharmacology*, 57 (12), pp. 891–896.
[253] Karousatos, C.M., Lee, J.K., Braxton, D.R. & Fong, T.L. (2021) Case series and review of Ayurvedic medication induced liver injury, *BMC Complementary Medicine & Therapies*, 21 (1), 91.
[254] Sridharan, K., Mohan, R., Ramaratnam, S. & Panneerselvam, D. (2011) Ayurvedic treatments for diabetes mellitus, *Cochrane Database Systematic*

Review, 12, CD008288.
[255] Bannuru, R.R., Osani, M.C., Al-Eid, F. & Wang, C. (2018) Efficacy of curcumin and Boswellia for knee osteoarthritis: Systematic review and meta-analysis, *Seminars in Arthritis & Rheumatism*, 48 (3), pp. 416–429.
[256] https://www.ncbi.nlm.nih.gov/pmc/articles/PMC2605614/
[257] Coon, J.T. & Ernst, E. (2004) Andrographis paniculata in the treatment of upper respiratory tract infections: A systematic review of safety and efficacy, *Planta Medica*, 70 (4), pp. 293–298.
[258] Paudyal, P., Jones, C., Grindey, C., Dawood, R. & Smith, H. (2018) Meditation for asthma: Systematic review and meta-analysis, *Journal of Asthma*, 55 (7), pp. 771–778.
[259] Salhofer, I., Will, A., Monsef, I. & Skoetz, N. (2016) Meditation for adults with haematological malignancies, *Cochrane Database Systematic Review*, 2 (2), CD011157.
[260] Goyal, M., Singh, S., Sibinga, E.M.S., Gould, N.F., Rowland-Seymour, A., Sharma, R., Berger, Z., Sleicher, D., Maron, D.D., Shihab, H.M., Ranasinghe, P.D., Linn, S., Saha, S., Bass, E.B. & Haythornthwaite, J.A. (2014) Meditation programs for psychological stress and well-being, *Agency for Healthcare Research and Quality (US)*, Report No. 13(14)-EHC116-EF.
[261] Krisanaprakornkit, T., Ngamjarus, C., Witoonchart, C. & Piyavhatkul, N. (2010) Meditation therapies for attention-deficit/hyperactivity disorder (ADHD), *Cochrane Database Systematic Review*, 2010 (6), CD006507.
[262] Gard, T., Hölzel, B.K. & Lazar, S.W. (2014) The potential effects of meditation on age-related cognitive decline: A systematic review, *Annals of the New York Academy of Sciences*, 1307, pp. 89–103.
[263] Ernst, E. & Lee, M.S. (2010) *Focus on Alternative and Complementary Therapies*, 15 (4), pp. 274–327.
[264] Matsushita, T. & Oka, T. (2015) A large-scale survey of adverse events experienced in yoga classes, *Biopsychosocial Medicine*, 18, 9.
[265] https://www.social-consciousness.com/2017/06/vaticans-chief-exorcist-warns-that-yoga-causes-demonic-possession.html
[266] https://www.theguardian.com/lifeandstyle/2020/jun/26/experience-my-yoga-class-turned-out-to-be-a-cult
[267] https://www.karmacentre.co.uk/ayurveda-clinic/
[268] Fox, M., Dickens, A., Greaves, C., Dixon, M. & James, M. (2006) Marma therapy for stroke rehabilitation—a pilot study, *Journal of Rehabilitation Medicine*, 38 (4), pp. 268–271.
[269] Brandling, J. & House, W. (2009) Social prescribing in general practice: Adding meaning to medicine, *British Journal of General Practice*, 59, pp. 454–456.
[270] https://www.healthline.com/health/social-prescribing#what-is-it
[271] HRH The Prince of Wales (2021) A message from HRH The Prince of Wales, honorary fellow of the Royal College of Physicians, *Future Healthcare Journal*, 8 (1), pp. 5–7.

272 Schmidt, K. & Ernst, E. (2003) MMR vaccination advice over the internet, *Vaccine*, 21 (11–12), pp. 1044–1047.
273 https://www.investopedia.com/terms/b/bait-switch.asp
274 Wood, E., Ohlsen, S., Fenton, S.J., Connell, J. & Weich, S. (2021) Social prescribing for people with complex needs: A realist evaluation, *BMC Family Practice*, 22 (1), p. 53.
275 Dixon, M. & Ornish, D. (2021) Love in the time of COVID-19: Social prescribing and the paradox of isolation, *Future Healthcare Journal*, 8 (1), pp. 53–56.
276 https://www.england.nhs.uk/personalisedcare/social-prescribing/
277 Wood, E., Ohlsen, S., Fenton, S.J., Connell, J. & Weich, S. (2021) Social prescribing for people with complex needs: A realist evaluation, *BMC Family Practice*, 22 (1), p. 53.
278 https://www.nhs.uk/conditions/homeopathy/
279 https://edzardernst.com/2018/01/homeopathy-for-farmers-with-the-support-of-prince-charles-foolish-and-immoral/
280 https://www.ainsworths.com/
281 https://edzardernst.com/2013/10/a-fictitious-interview-with-the-prince-of-wales/
282 https://collegeofmedicine.org.uk/homeopathy-theories-of-mechanisms/
283 https://pressreleases.responsesource.com/news/97940/hrh-prince-charles-announced-as-new-patron-of-the-faculty/
284 https://edzardernst.com/2014/06/berlin-wall-homeopathy-at-its-finest/
285 https://www.hive.co.uk/Product/Edzard-Ernst/Homeopathy---The-Undiluted-Facts--Including-a-Comprehensive-A-Z-Lexicon/19719982
286 Stolberg, M. (2006) Inventing the randomized double-blind trial: The Nuremberg salt test of 1835, *Journal of the Royal Society of Medicine*, 99 (12), pp. 642–643.
287 Ernst, E. (2010) Homeopathy: What does the 'best' evidence tell us?, *Medical Journal of Australia*, 192 (8), pp. 458–460.
288 Brien, S., Lachance, L., Prescott, P., McDermott, C. & Lewith, G. (2011) Homeopathy has clinical benefits in rheumatoid arthritis patients that are attributable to the consultation process but not the homeopathic remedy: A randomized controlled clinical trial, *Rheumatology (Oxford)*, 50 (6), pp. 1070–1082.
289 Mathie, R.T. & Clausen, J. (2015) Veterinary homeopathy: Meta-analysis of randomised placebo-controlled trials, *Homeopathy*, 104 (1), pp. 3–8.
290 Doehring, C. & Sundrum, A. (2016) Efficacy of homeopathy in livestock according to peer-reviewed publications from 1981 to 2014, *Veterinary Record*, 179 (24), p. 628.
291 https://edzardernst.com/2018/03/homeopathy-for-cancer-dr-wurster-and-the-clinica-sta-croce-in-switzerland/
292 https://edzardernst.com/2019/10/vidatox-homeopathys-answer-to-cancer-or-outright-fraud/
293 https://www.thetimes.co.uk/article/medical-science-can-find-no-cure-

for-the-belief-in-quack-remedies-5s5hxpflh
294 https://www.deccanherald.com/national/covid-19-bengaluru-specialist-cured-prince-charles-claims-ayush-minister-clarence-house-denies-820546.html
295 https://www.amazon.co.uk/Rebel-Prince-Defiance-explosive-biography-ebook/dp/B079172F2M/ref=sr_1_1?dchild=1&keywords=bower%2C+prince+charles&qid=1632386163&s=books&sr=1-1
296 Ernst, E. & Zhang, J. (2011) Cardiac tamponade caused by acupuncture: A review of the literature, *International Journal of Cardiology*, 149 (3), pp. 287–289.
297 Posadzki, P., Watson, L.K. & Ernst, E. (2013) Adverse effects of herbal medicines: An overview of systematic reviews, *Clinical Medicine (London)*, 13 (1), pp. 7–12.
298 Ernst, E. (2010) Assessments of complementary and alternative medicine: The clinical guidelines from NICE, *International Journal of Clinical Practice*, 64 (10), pp. 1350–1358.
299 https://www.bbc.com/news/uk-32740154
300 https://www.england.nhs.uk/2018/06/nhs-england-welcomes-homeopathy-court-ruling/
301 https://www.politicmag.net/politics-news/prince-charles-rejected-by-experts-before-gwyneth-paltrows-long-covid-row-witchcraft-royal-news-reports-40164-2021/
302 https://www.princes-trust.org.uk/
303 https://www.goodhousekeeping.com/uk/news/a24839545/prince-charles-role-monarch/
304 https://www.express.co.uk/news/royal/1402457/prince-charles-coronavirus-gwyneth-paltrow-covid-19-nhs-royal-family-spt
305 https://www.rt.com/uk/416279-prince-charles-alternative-medicine/
306 https://www.telegraph.co.uk/news/uknews/2700298/Prince-Charles-therapist-sued-by-amputee.html
307 https://www.independent.co.uk/news/uk/politics/prince-charles-lobbied-prime-minister-support-alternative-medicines-letters-show-10247842.html
308 https://www.msn.com/en-gb/news/uknews/king-charles-iii-will-carry-on-championing-green-issues/ar-AA11UKWy?li=BBoPRmx
309 https://www.the-scientist.com/news-opinion/the-unscientific-king-charles-iii-s-history-promoting-homeopathy-70544
310 https://edition.cnn.com/2022/09/22/uk/future-of-the-monarchy-king-charles-queen-elizabeth-intl-cmd-gbr/
311 https://uk.finance.yahoo.com/news/cash-honours-probe-launched-charles-122025452.html?guccounter=1
312 https://nypost.com/2022/06/25/prince-charles-took-millions-in-cash-from-qatari-sheikh-report/
313 https://en.wikipedia.org/wiki/Prince_Andrew,_Duke_of_York
314 https://www.theguardian.com/uk-news/2023/jan/06/prince-harry-saw-

red-mist-in-william-during-alleged-attack
315 https://www.theguardian.com/commentisfree/2022/sep/08/king-charles-iii-monarchy-mother-nation
316 Hawke, K., King, D., van Driel, M.L. & McGuire, T.M. (2022) Homeopathic medicinal products for preventing and treating acute respiratory tract infections in children, *The Cochrane Database of Systematic Reviews*, 12 (12), CD005974.
317 Alshami, A. & Alqassab, F.H. (2022) The short-term effects of instrument-based mobilization compared with manual mobilization for low back pain: A randomized clinical trial, *Journal of Back and Musculoskeletal Rehabilitation*, 10.3233. Advance online publication: https://doi.org/10.3233/BMR-220042
318 Naranjo-Cinto, F., Cerón-Cordero, A.I., Figueroa-Padilla, C., Galindo-Paz, D., Fernández-Carnero, S., Gallego-Izquierdo, T., Nuñez-Nagy, S. & Pecos-Martín, D. (2022) Real versus sham manual therapy in addition to therapeutic exercise in the treatment of non-specific shoulder pain: A randomized controlled trial, *Journal of Clinical Medicine*, 11 (15), 4395.
319 Hauptmann, M., Kutschan, S., Hübner, J. & Dörfler, J. (2022) Bioenergy therapies as a complementary treatment: A systematic review to evaluate the efficacy of bioenergy therapies in relieving treatment toxicities in patients with cancer, *Journal of Cancer Research and Clinical Oncology*, 10.1007. Advance online publication: https://doi.org/10.1007/s00432-022-04362-x
320 Mukerji, N. & Ernst, E. (2022) Why homoeopathy is pseudoscience, *Synthese*, 200, 394.
321 Posadzki, P., Kyaw, B.M., Dziedzic, A. & Ernst, E. (2022) Osteopathic manipulative treatment for pediatric conditions: An update of systematic review and meta-analysis, *Journal of Clinical Medicine*, 11 (15), 4455.
322 Vezzetti, A., Rosati, L.M., Lowe, F.J., Graham, C.B., Moftakhar, R., Mangubat, E. & Sen, S. (2022) Stenting as a treatment for cranio-cervical artery dissection: Improved major adverse cardiovascular event-free survival, *Catheterization and Cardiovascular Interventions*, 99 (1), pp. 134–139.
323 Kuyken, W., *et al.* (2022) Effectiveness and cost-effectiveness of universal school-based mindfulness training compared with normal school provision in reducing risk of mental health problems and promoting well-being in adolescence: The MYRIAD cluster randomised controlled trial, *Evidence-Based Mental Health*, 25 (3), pp. 99–109. Advance online publication: https://doi.org/10.1136/ebmental-2021-300396
324 Egli, M., Deforth, M., Keiser, S., Meyenberger, P., Muff, S., Witt, C.M. & Barth, J. (2022) Effectiveness of a brief hypnotic induction in third molar extraction: A randomized controlled trial (HypMol), *The Journal of Pain*, 23 (6), pp. 1071–1081.
325 Seifert, G., *et al.* (2022) Integrative medicine during the intensive phase of chemotherapy in pediatric oncology in Germany: A randomized controlled trial with 5-year follow up, *BMC Cancer*, 22 (1), 652.

[326] White C.M. (2022) Continued risk of dietary supplements adulterated with approved and unapproved drugs: Assessment of the US Food and Drug Administration's Tainted Supplements Database 2007 through 2021, *Journal of Clinical Pharmacology*, 62 (8), pp. 928–934.

[327] Prely, H., Herledan, C., Caffin, A.G., Baudouin, A., Larbre, V., Maire, M., Schwiertz, V., Vantard, N., Ranchon, F. & Rioufol, C. (2022) Real-life drug-drug and herb-drug interactions in outpatients taking oral anticancer drugs: Comparison with databases, *Journal of Cancer Research and Clinical Oncology*, 148 (3), pp. 707–718.

[328] Wagenknecht, A., Dörfler, J., Freuding, M., Josfeld, L. & Huebner, J. (2022) Homeopathy effects in patients during oncological treatment: A systematic review, *Journal of Cancer Research and Clinical Oncology*, 10.1007. Advance online publication: https://doi.org/10.1007/s00432-022-04054-6

[329] Perkins, P., Parkinson, A., Parker, R., Blaken, A. & Akyea, R.K. (2022) Does acupressure help reduce nausea and vomiting in palliative care patients? A double blind randomised controlled trial, *BMJ Supportive & Palliative Care*, 12 (1), pp. 58–63.

[330] Adler, U.C., Adler, M.S., Padula, A.E.M., Hotta, L.M., de Toledo Cesar, A., Diniz, J.N.M., de Freitas Santos, H. & Martinez, E.Z. (2022) Homeopathy for COVID-19 in primary care: A randomized, double-blind, placebo-controlled trial (COVID-Simile study), *Journal of Integrative Medicine*, 20 (3), pp. 221–229.

[331] Loeb, M., Russell, M.L., Neupane, B., Thanabalan, V., Singh, P., Newton, J. & Pullenayegum, E. (2018) A randomized, blinded, placebo-controlled trial comparing antibody responses to homeopathic and conventional vaccines in university students, *Vaccine*, 36 (48), pp. 7423–7429.

[332] Raak, C.K., Scharbrodt, W., Berger, B., Büssing, A., Schönenberg-Tu, A., Martin, D.D., Robens, S. & Ostermann, T. (2022) Hypericum perforatum to Improve Postoperative Pain Outcome After Monosegmental Spinal Sequestrectomy (HYPOS): Results of a randomized, double-blind, placebo-controlled trial, *Journal of Integrative and Complementary Medicine*, 28 (5), pp. 407–417.

[333] Rai, S., Gupta, G.N., Singh, S., Michael, J., Misra, P., Gupta, B., Singh, S., Prakash, A., Tomar, M., Sadhukhan, S., Koley, M. & Saha, S. (2022) Efficacy of individualized homeopathic medicines in treatment of acne vulgaris: A double-blind, randomized, placebo-controlled trial, *Homeopathy: The Journal of the Faculty of Homeopathy*, 111 (4), pp. 240–251.

[334] Gartlehner, G., Emprechtinger, R., Hackl, M., Jutz, F.L., Gartlehner, J.E., Nonninger, J.N., Klerings, I. & Dobrescu, A.I. (2022) Assessing the magnitude of reporting bias in trials of homeopathy: A cross-sectional study and meta-analysis, *BMJ Evidence-Based Medicine*, 27 (6), pp. 345–351.

[335] de Albuquerque, L.M.N.F. & Turrini, R.N.T. (2022) Effects of flower essences on nursing students' stress symptoms: A randomized clinical trial, *Revista da Escola de Enfermagem da U S P*, 56, e20210307. https://doi.org/10.1590/1980-220X-REEUSP-2021-0307

Index

Act of Parliament 39, 44
Acupressure 183
Acupuncture 7, 31, 51, 53, 62, 67, 90, 93, 96, 99, 100, 136, 154, 155, 174
Adverse effect 38, 46, 47, 48, 79, 87, 93, 94, 97, 152
Advisor, advice 28, 39, 55, 78, 79, 178
Age of reason 126–31, 166
Alchemy 24, 25, 27
Ali, Mosaraf 62–65, 132, 153
Anthroposophy 121, 122, 164
Antibiotic 142–46, 154, 156, 165, 176
Architecture 5, 119, 130, 132
Armstrong, Boo 104, 105
Artery dissection 182
Ayurveda 62, 136, 138, 147–53

Back pain 40, 41, 43, 46, 85, 92
Bacteria 142–46
Baker, Andrew 110, 112
Balance 26, 38, 39, 90, 133, 140, 147
Bathija, Raj 64
Baum, Michael 74, 75, 86
BBC 17, 100

Bedell Smith, Sally 15
Bioenergy 182
Black spider memos 115–19
Blair, Tony 78
Blood 38, 39, 135
Bower, Tom 27
Bravewell Collaborative 106–09, 178
British Chiropractic Association 43, 48
British Diabetic Associates 113
British Medical Association, BMA 25–33, 50
British Medical Journal, BMJ 30, 66, 74
British School of Osteopathy 38, 39
Bushman 17, 20, 23

Cancer 5, 23, 70, 73–76
Cash for honours 181
Chandler, Cyril 124
Charity 53, 54
Chemotherapy 74, 136
Chiropractic 7, 31, 38, 43–49, 78, 161, 174
Clinical trial 33, 59, 72, 163
Cochrane Collaboration 4, 85,

136, 137, 150
Cohen, Nick 134
College of Medicine and Integrated Health 55, 120–25, 147, 149, 161, 181
Colquhoun, David 80, 87, 104, 120, 121, 126
Compassion 67, 69, 120, 164, 175
Complementary and Natural Healthcare Council, CNHC 51, 53
Cost 40, 53, 54, 73, 82, 83, 84, 89, 94, 117, 166
Cost effectiveness 182
COVID 149, 167
Crichton-Miller, Mrs. 22
Critical analysis/critical thinking 2, 3, 11, 32, 127, 129, 130
Crossfields Institute 121

Death 6, 18, 20, 46, 47, 49, 76, 87, 139, 144
Department of Health, DoH 53, 55, 86, 101, 123
Depression/diagnostic 77, 81, 90, 92, 116, 117, 124, 149, 154
Detox 70, 73, 110–14, 147
Diagnosis 5, 23, 37, 42, 45, 46, 63, 64, 77, 98, 137, 139, 170
Dietary supplement 46, 73, 110, 112, 124, 183
Dimbleby, Jonathan 14, 16, 18, 25, 27, 43, 161
Dixon, Michael 52, 55, 79, 120–25, 132, 147, 148, 158
Drug 27, 28, 38, 45, 46, 80, 85, 90, 97, 98, 107, 111, 120, 142
Duke of Sussex 181

Dutchy Originals 110–14, 177

Effectiveness/efficacy 7, 40, 41, 46, 52, 59, 68, 69, 75, 82, 87, 94, 117, 131, 136, 157, 170, 176
Empathy 39, 67, 69, 164, 175
Empiricism 127, 134
Energy 16, 163
Enlightenment 126–31
Environment 180
Establishment 27, 28, 29, 32, 107, 115, 152, 169, 170, 171, 177
Ethics 46, 95, 117, 166
European Academies' Science Advisory Council 98
Evidence 2, 3, 4, 6, 29, 34, 35, 36, 39, 40, 41, 46, 47, 48, 51, 52, 54, 61, 63, 71, 75, 79, 80, 82, 84, 86, 89, 91, 92, 97, 98, 99, 114, 116, 122, 123, 127, 129, 135, 137, 139, 147, 150, 157, 162, 166, 176
Evidence-based medicine, EBM 70, 72, 116, 145, 164, 166
Exeter 7, 8, 35, 52, 64, 83, 153
Experience 4, 127, 129, 133
Expert 6, 28, 35, 59, 72, 83, 86, 129, 139, 164, 165, 170, 179

Faculty of Homeopathy 14, 161, 181
Fallacy 127, 156, 158, 173
Farming 142, 144, 145, 161, 162
Federation of European Academies of Medicine, FEAM 98
Fielding, Simon 121, 125
Flower essence 184

Flynn, Paul 118
Fraud 53, 104
Freud, Sigmund 24

General Medical Council 30
General practitioner, GP 33, 52, 69, 85, 100, 102, 103, 124, 148, 154
General Chiropractic Council 44, 48
General Medical Council 63, 84
General Osteopathic Council 38, 39, 121
Gerson therapy 5, 6, 73–76
GetWellUK 100–05, 177
Giffard Ingaret 17, 22
Gorski, David 107, 108
Government 101, 105, 119
Gray, George 54

Hahnemann, Samuel 11, 14, 16, 160, 164
Hain, Peter 100, 104
Hall, Harriet 109
Harm 7, 35, 47, 52, 54, 59, 75, 77, 95, 128, 154, 157, 165, 179
Harmony 1, 12, 14, 23, 39, 101, 127, 132–41, 147, 162, 172, 175
Hastings, Charles 30
Health Secretary 79, 115, 118, 170
Herb drug interaction 183
Herbal medicine 7, 31, 43, 44, 51, 52, 53, 77–81, 90, 93, 96, 110, 112, 113, 155, 174
Hess, Rudolf 62, 65, 68
Himmler, Heinrich 65
Hitchens, Christopher 23
Holism 12, 33, 59, 68, 69, 108, 112, 124, 175
Homeopathy 6, 7, 11, 14, 16, 23, 27, 31, 39, 44, 53, 62, 65, 68, 84, 85, 86, 110, 116, 117, 131, 134, 136, 142, 143, 144, 145, 160–68, 170, 174, 175, 177, 180, 182, 183, 184
Horton, Richard 84
Hospital 6, 14, 62–65, 73, 116, 161, 177
House of Lords 59, 61, 105
Huston, John 16
Hutchinson, Ian 43
Hypericum 183
Hypnosis 63, 122, 183
Hypothesis 4, 59, 94, 164

Informed consent 46, 95
Integrated or integrative medicine/healthcare 12, 50–55, 62, 63, 64, 67–72, 73, 78, 83, 91, 101, 105, 106, 107, 112, 120–25, 131, 145, 156, 158, 176, 183
Intuition/instinct 1, 2, 16, 32, 36, 90, 127, 135, 166, 169, 171
Ioannidis, John 7
Iridology 5, 63, 137, 138, 170

Jeffrey Epstein 181
Jinping, Xi 96
Johnson, Alan 115, 170
Jones, J.D.F. 17, 22, 23
Journalist 4, 6, 119
Jung, Carl 17, 19, 20, 24
Juniper, Tony 132, 134, 140

Kalahari Desert 17, 19, 23

King's Fund 44, 45
Kohler-Baker, Bonnie 22

Lady Diana 13, 17, 20
Laing, Maurice 35
Lavely, Kim 73
Lillard, Harvey 49
Littlejohn, John Martin 38
Lord Kindersley 43

Mack, Christy 106, 107
Mack, John 107
Manipulation 38, 40, 43, 45, 47, 48, 67, 85, 154
Manual therapy 37, 47, 182
Marma therapy 63, 65, 153
Massage 38, 67, 147, 148, 154
Mayer, Catherine 6, 28
McCabe, Steve 118
Meditation 147, 150, 151
Medline 27, 93
Mental health 102, 151, 154
Mills, Simon 124, 125
Mindfulness 182
Mobilisation 38
Modi, Narendra 147, 149
Mostert, Cari 20
Mysticism 15, 16, 20, 134

Naik, Shripad 167
National Health Service, NHS 50, 51, 61, 81, 82, 86, 87, 88, 99, 100, 101, 104, 105, 116, 118, 120, 148, 156, 159, 160, 161, 166, 176, 179
Nature, natural 12, 15, 19, 26, 28, 37, 63, 68, 77, 96, 110, 112, 132, 135, 147, 160, 163, 170, 172

Naturopathy 31, 111, 148
Novella, Steven 113

Osteopathy 31, 37–44, 78, 100, 121, 127, 135, 136, 160

Palmer, David Daniel 43, 48
Paracelsus 25, 27, 29
Patron 14, 39, 55, 120, 127, 161, 165
Peat, Michael 83
Physiology 37, 39, 97, 137, 138, 139
Phytotherapy 77, 79, 81, 178
Placebo 4, 46, 61, 85, 104, 128, 143, 144, 160, 162, 164, 165
Plausibility 40, 93, 137, 139, 144, 160, 164, 170, 178
Politics 1, 2, 45, 82, 83, 86, 105, 115–19, 177
Porter, Shirley 86
Prevention 38, 39, 40, 92, 145
Prince Andrew 181
Prince Philip 14
Prince's Trust 13, 177
Princess Anne 39
Progress 2, 32
Pseudoscience 33, 40, 52, 105, 108, 123, 131, 134
Psychotherapy 17, 24, 169
Public health 2, 70, 71, 158
Pukka herbs 123
Pulse diagnosis 138

Quackery 70, 71, 84, 104, 112, 129, 134, 140, 162
Queen Elizabeth 14, 45, 117
Quin, Frederic Hervey Foster 14

Rationality 55, 80, 84, 130, 135, 145, 179
Read, David 94
Reflexology 59, 137, 138, 139, 148
Regulation 41, 43, 44, 48, 51, 53, 78, 81, 176
Religion 16, 49, 129, 150
Renaissance 19, 25, 181
Research 33, 41, 47, 50, 51, 53, 54, 59, 74, 81, 88, 93, 94, 99, 107, 109, 131, 178
Risk 7, 46, 48, 61, 69, 79, 125, 163
Royal College of Chiropractors, RCC 44, 47
Royal Society of Medicine, RSM 31–35

Safety 7, 50, 93, 98, 99, 125, 131, 157, 164, 165, 170
Sallis, Zoe 15, 16
Sceptic 130, 140, 166, 170
Side effect 41, 90
Singh, Simon 48, 143
Skelly, Ian 132
Skrabanek, Petr 129
Smallwood, Christopher 82, 92
Smallwood report 82–88, 94, 161, 177
Smith, Steve 88
Social prescribing 125, 155–59, 175
Spine 38, 40, 43, 45
Spiritualism 16, 19, 23, 26, 39, 48, 55, 87, 134, 137, 169
Steiner, Rudolf 121
Still, Andrew Taylor 37, 39, 42
St John's Wort 77, 81, 90, 92

Student 30, 54
Subluxation 43, 44, 49
Surgeon/surgery 31, 38, 42
Survey 59, 152
Systematic review 4, 40, 72, 85, 86, 122, 136, 151, 152, 164, 165

Thought field therapy 122, 123
Tongue diagnosis 138
Tooke, John 52
Traditional Chinese Medicine, TCM 94, 96–99, 138
Traditional wisdom 26, 66, 90, 172

Unani medicine 63
University College of Osteopathy 39, 41

Vaccination 46, 49, 70
Van der Post Laurens 16–24, 25, 27, 169, 180
Vitalism/vital force 16, 49, 97, 163

Waitrose 110
Watt, James 31, 35, 44
Weil, Andrew 107, 109, 125
Well-being 38, 87, 100, 102, 154
World Health Organisation, WHO 77, 83, 89, 90–95

Ying, Fu 96
Yoga 63, 147–53, 175

Zedong, Mao 96

www.ingramcontent.com/pod-product-compliance
Lightning Source LLC
Chambersburg PA
CBHW070942230426
43666CB00011B/2531